ESD BASICS

ESD Series
By Steven H. Voldman

ESD Basics: From Semiconductor Manufacturing to Product Use
ISBN: 9780470979716
September 2012

ESD: Design and Synthesis
ISBN: 9780470685716
March 2011

ESD: Failure Mechanisms and Models
ISBN: 9780470511374
July 2009

Latchup
ISBN: 9780470016428
December 2007

ESD: RF Technology and Circuits
ISBN: 9780470847558
September 2006

ESD: Circuits and Devices
ISBN: 9780470847541
November 2005

ESD: Physics and Devices
ISBN: 9780470847534
September 2004

Upcoming titles:

ESD: Test and Characterization

Electrical Overstress (EOS): Devices, Circuits, and Systems

The ESD Handbook

ESD BASICS
From Semiconductor
Manufacturing to Product Use

Steven H. Voldman
IEEE Fellow, Vermont, USA

A John Wiley & Sons, Ltd., Publication

Library of Congress Cataloging-in-Publication Data
Voldman, Steven H.
   ESD basics : from semiconductor manufacturing to product use / Steven Voldman.
      p. cm.
   Includes bibliographical references and index.
   ISBN 978-0-470-97971-6 (hardback)
   1. Electronic apparatus and appliances–Design and construction.   2. Electric discharges.   3. Electronic apparatus and appliances–Protection.   4. Microelectronics.   5. Static eliminators.   6. Electrostatics.   I. Title.
   TK7836.V65 2012
   621.3815–dc23
                                                                                        2012018125

A catalogue record for this book is available from the British Library.

Print ISBN: 9780470979716

Set in 10/12pt, Times by Thomson Digital, Noida, India.

To My Parents
Carl and Blossom Voldman

Contents

About the Author

Dr Steven H. Voldman is the first IEEE Fellow in the field of electrostatic discharge (ESD) for "Contributions in ESD protection in CMOS, Silicon On Insulator and Silicon Germanium Technology." He received his B.S. in Engineering Science from University of Buffalo (1979); a first M.S. EE (1981) from Massachusetts Institute of Technology (MIT); a second degree EE Degree (Engineer Degree) from MIT; a MS Engineering Physics (1986) and a PhD in Electrical Engineering (EE) (1991) from University of Vermont under IBM's Resident Study Fellow program.

He was a member of the IBM development for 25 years working on semiconductor device physics, device design, and reliability (e.g., soft error rate (SER), hot electrons, leakage mechanisms, latchup and electrostatic discharge (ESD)). Steve Voldman has been involved in latchup technology development for 27 years. He worked on both technology and with product development in Bipolar SRAM, CMOS DRAM, CMOS logic, Silicon on Insulator (SOI), BiCMOS, Silicon Germanium (SiGe), RF CMOS, RF SOI, smart power, and image processing technologies. In 2008, he was a member of the Qimonda DRAM development team, working on 70, 58 and 48 nm CMOS technology. In 2008, he initiated a limited liability corporation (LLC), and he worked at headquarters in Hsinchu, Taiwan for Taiwan Semiconductor Manufacturing Corportion (TSMC) as part of the 45 nm ESD and latchup development team. He was a Senior Principal Engineer working for the Intersil Corporation on ESD and latchup development from 2009 to 2011. Since 2011, he is presently independent under Dr Steven H. Voldman LLC providing consulting, teaching, and patent litigation expert witness support.

Dr Voldman was chairman of the SEMATECH ESD Working Group from 1995 to 2000. In his SEMATECH Working Group, the effort focused on ESD technology benchmarking, the first transmission line pulse (TLP) standard development team, strategic planning, and JEDEC-ESD Association standards harmonization of the human body model (HBM) standard. From 2000 to 2010, as Chairman of the ESD Association Work Group on TLP and very-fast TLP (VF-TLP), his team was responsible for initiating the first standard practice and standards for TLP and VF-TLP. Steve Voldman has been a member of the ESD Association Board of Directors, and Education Committee. He initiated the "ESD on Campus" program which was established to bring ESD lectures and interaction to university faculty and students internationally; the ESD on Campus program has reached over 40 universities in the United States, Singapore, Taiwan, Malaysia, Philippines, Thailand, India, and China.

Dr Voldman teaches short courses and tutorials on ESD, latchup, and invention in the United States, China, Singapore, Malaysia, Taiwan, Sri Lanka, and Israel. He is a recipient of over 240 issued US patents, in the area of ESD and CMOS latchup. He has served as an expert witness in patent litigation cases associated with ESD and latchup.

Dr Voldman has also written articles for *Scientific American* and is an author of the first book series on ESD and latchup: *ESD: Physics and Devices*, *ESD: Circuits and Devices*, *ESD*: *RF Technology and Circuits, Latchup, ESD: Failure Mechanisms and Models*, and *ESD: Design and Synthesis* as well as a contributor to the books *Silicon Germanium: Technology, Modeling and Design* and *Nanoelectronics: Nanowires, Molecular Electronics, and Nano-devices*. In addition, the International Chinese editions of book *ESD: Circuits and Devices* and the text *ESD*: *RF Technology and Circuit* is also released.

Preface

This text, *ESD Basics: From Semiconductor Manufacturing to Product Use* was initiated on the need to produce a text that addresses fundamentals of electrostatic discharge from the manufacturing environment to today's products. As the manufacturing world evolves, semiconductor networks scale, and systems are changing, the needs and requirements for reliability and ESD protection are changing. A text is required that connects basic ESD phenomena to today's real world environment.

Whereas significant texts are available today to teach experts on ESD on-chip design, there is a need for non-experts, non-technical and layman to understand the problems facing the world today. Today, real world ESD and EMI issues surround us; from exploding gas tanks, electrostatic discharge problems in automobiles, cable-induced latchup of computer servers, to automotive noise related issues. Hence, there is a need for non-experts to understand the issues that revolve around us, and what do we do to avoid them.

This text has multiple goals.

The first goal of the text is to teach the basics of electrostatics, and tribo-electric charging and relate them to the electrostatic discharge processes in semiconductor manufacturing, handling and assembly. While teaching some fundamentals, I added some history as well.

The second goal of the text is to teach the electrostatic discharge (ESD), tribo-charging, electrical overstress (EOS), and latchup subject matter. The text opens the door on the issues of electromagnetic interference (EMI) and electromagnetic compatibility (EMC).

The third goal is to relate these processes to modern day chips, and systems in today's world. Hopefully, the examples of problems today will make it more relevant, entertaining and a little fun as well.

The fourth goal is to demonstrate how to protect semiconductor chips with on-chip protection networks.

The fifth goal is to expose the reader to ESD testing of both semiconductor chips and systems.

The sixth goal is to discuss where the future lies with ESD phenomena, standards, testing and future products.

The seventh goal is to provide a glimpse into the future with new nano-structures and nano-systems, and the anticipated electrostatic and electromagnetic issues ahead!

This text, *ESD: From Semiconductor Manufacturing to Product Use* will contain the following:

- Chapter 1 introduces the reader to an overview of the language and fundamentals associated with electrostatics. In Chapter 1, a brief discussion of electrostatics, and tribo-electrical phenomena is weaved into the individuals, dates, and history: Thales of Miletus, Gray,

Dufay, Franklin, Toepler, Faraday, Cavendish, Coulomb, to Maxwell – just to mention a few. The chapter then quickly fast-forwards to today's issues of electrostatic discharge (ESD), electrical overstress (EOS), latchup, electromagnetic interference (EMI), and electromagnetic compatibility (EMC) in components and systems.

- Chapter 2 discusses electrostatic discharge control in manufacturing environments. In this chapter, the objective is to provide the reader with a taste of the issues, test methods, standards, and control programs in manufacturing to provide an ESD protected area.

- In Chapter 3, the subject switches to focus on a deeper look at electrostatic discharge (ESD), electrical overstress (EOS), electromagnetic interference (EMI) and electromagnetic compatibility (EMC). Each of these fields has a vast number of publications, literature, and books. In the introduction, I have provided some of the language, terms, and testing standards.

- In Chapter 4, system level concerns associated with ESD, EOS, latchup, EMI, and EMC have been discussed relevant to today's application and the future. A brief discussion of electrostatic issues in servers, laptops, hand-held devices, cell phones, disk drives, digital cameras, autos, and space applications was reviewed to educate the reader in the vast number of issues in today's electronic environment. System level ESD tests, such as IEC 61000-4-2, HMM, CDE, and CBM are discussed.

- Chapter 5 focuses on semiconductor component solutions. The focus of the chapter is on "on-chip" ESD protection networks. ESD protection in digital, analog, and RF applications are discussed. In this chapter, ESD circuits schematics, layout, and semiconductor chip floor planning are also discussed.

- Chapter 6 focuses on system level solutions. System level solutions being practiced today are shown for ESD, EMI, and EMC. New concepts such as system level EMC scanning techniques are discussed.

- Chapter 7 discusses ESD protection for today's and tomorrow's nanostructure technology. As the dimensional scaling of devices approaches the nano-meter dimensions, all devices will have to address the implications of static charge, electrostatic discharge (ESD), electromagnetic interference (EMI), and electrical overstress (EOS). This will be true in photo-masks, magnetic recording devices, semiconductor devices, nano-wires to nano-tubes. This concluding chapter takes a look at micro-motors, micro-mirrors, RF MEM switches, and many novel devices.

This introductory text will hopefully pique your interest in the field of electrostatic discharge (ESD), electrical overstress (EOS), electromagnetic interference (EMI), and electromagnetic compatibility (EMC) – and teach how it relates to today's world. To establish a stronger knowledge of ESD protection, it is advisable to read the other texts *ESD: Physics and Devices, ESD: Circuits and Technology, ESD: RF Circuits and Technology, ESD: Failure Mechanisms and Models, ESD: Design and Synthesis*, and *Latchup*.

Enjoy the text, and enjoy the subject of ESD, EOS, latchup, EMI, and EMC phenomena.

Baruch HaShem (B"H)

Dr Steven H. Voldman
IEEE Fellow

Acknowledgments

I would like to thank the years of support from the SEMATECH, the ESD Association, the IEEE, and the JEDEC organizations. I would like to thank the IBM Corporation, Qimonda, Taiwan Semiconductor Manufacturing Corporation (TSMC), and the Intersil Corporation. The work in this text comes from thirty years of working with bipolar memory, DRAM memory, SRAM, NVRAMs, microprocessors, ASICs, mixed voltage, mixed signal, RF to power applications. I was fortunate to work in a wide number of technology teams, and with a wide breadth of customers. I was very fortunate to work in bipolar memory, CMOS DRAM, CMOS logic, ASICs, silicon on insulator (SOI), and Silicon Germanium (SiGe) from 1 μm to 45 nm technologies. I was very fortunate to be a member of talented technology and design teams that were both innovative, intelligent, and inventive. This provided opportunity to provide experimental concepts, and try new ideas in ESD design in applications and products.

I would like to thank the institutions that allowed me to teach and lecture at conferences, symposiums, industry, and universities; this gave me the motivation to develop the texts. I would like to thank faculty at the following universities: M.I.T., Stanford University, University of Central Florida (UCF), Vanderbilt University, University Illinois Urbana-Champaign (UIUC), University of California Riverside (UCR), University of Buffalo, National Chiao Tung University (NCTU), Tsin Hua University, National Technical University of Science and Technology (NTUST), National University of Singapore (NUS), Nanyang Technical University (NTU), Beijing University, Fudan University, Shanghai Jiao Tung University, Zheijang University, Huazhong University of Science and Technology (HUST), Universiti Sains Malaysia (USM), Universiti Putra Malaysia (UPM), Kolej Damansara Utama (KDU), Chulalongkorn University, Mahanakorn University, Kasetsart University, Thammasat University, and Mapua Institute of Technology.

I would like to say thank you for the years of support and the opportunity to provide lectures, invited talks, and tutorials to the *Electrical Overstress/Electrostatic Discharge (EOS/ESD) Symposium*, the *International Reliability Physics Symposium (IRPS)*, the *Taiwan Electrostatic Discharge Conference (T-ESDC)*, the *International Electron Device Meeting (IEDM)*, the *International Conference on Solid-State and Integrated Circuit Technology (ICSICT)*, the *International Physical and Failure Analysis (IPFA)*, *IEEE ASICON*, and *IEEE Intelligent Signal Processing And Communication Systems (ISPACS) Conference*.

I would like to thank my many friends for twenty years in the ESD profession – Prof. Ming Dou Ker, Prof. J.J. Liou, Prof. Albert Wang, Prof. Elyse Rosenbaum, Prof. Jo Chiranut Sa-ngiamsak, Timothy J. Maloney, Charvaka Duvvury, Eugene Worley, Robert Ashton, Yehuda Smooha, Vladislav Vashchenko, Ann Concannon, Albert Wallash, Vessilin Vassilev, Warren Anderson, Marie Denison, Alan Righter, Andrew Olney, Bruce Atwood, Jon Barth, Evan Grund, David Bennett, Tom Meuse, Michael Hopkins, Yoon Huh, Jin Min, Keichi Hasegawa, Nathan Peachey, Kathy Muhonen, Augusto Tazzoli, Gaudenzio Menneghesso, Marise BaFleur, Jeremy Smith, Nisha Ram, Swee K. Lau, Tom Diep, Lifang Lou, Stephen Beebe, Michael Chaine, Pee Ya Tan, Theo Smedes, Markus Mergens, Christian Russ, Harold Gossner, Wolfgang Stadler, Ming Hsiang Song, J. C. Tseng, J.H. Lee, Michael Wu, Erin Liao, Stephen Gaul, Jean-Michel Tschann, Tze Wee Chen, Shu Qing Cao, Slavica Malobabic, David Ellis, Blerina Aliaj, Lin Lin, David Swenson, Donn Bellmore, Ed Chase, Doug Smith, W. Greason, Stephen Halperin, Tom Albano, Ted Dangelmayer, Terry Welsher, John Kinnear, and Ron Gibson.

I would like to thank the ESD Association office for the support in the area of publications, standards developments, and conference activities. I would also like to thank the publisher and staff of John Wiley & Sons, for including this text as part of the ESD book series.

To my children, Aaron Samuel Voldman and Rachel Pesha Voldman, good luck to both of you in the future.

To my wife Annie Brown Voldman – thank you for the support and years of work.

And to my parents, Carl and Blossom Voldman.

Baruch HaShem (B"H)

Dr Steven H. Voldman
IEEE Fellow

1 Fundamentals of Electrostatics

1.1 INTRODUCTION

We are all familiar with electrostatic discharge (ESD): shuffle your feet across a shag carpet in your favorite sneakers, touch a piece of metal, and zap! For a human being, we let out an "ouch!"; but for micro-electronics to nano-electronics, this can lead to product failures [1].

But, today, and in the future, static charge will remain an important industrial issue for the production of both electronic devices to systems. It is also an issue in fields of munitions, explosives, chemical, and material industries. Any industry where there is a risk of impact to quality, yield, degradation, or physical harm will be concerned with electrostatic discharge (ESD), electrical overstress (EOS), electromagnetic interference (EMI), and electromagnetic compatibility (EMC).

In this book, a short survey of ESD from manufacturing to product use will be shown. The text will discuss fundamentals of electrostatics, manufacturing electrostatic issues, component level issues, system level issues, to design.

So, where did all this all begin?

1.2 ELECTROSTATICS

The discovery of electrostatic attraction and electrostatic discharge is one of the world's earliest understandings of scientific thought and analysis. Its first discovery goes back to the early foundation of the problem of the nature of matter, astronomy, mathematics and foundation of Greek philosophy, and pre-dates the nature of matter.

ESD Basics: From Semiconductor Manufacturing to Product Use, First Edition. Steven H. Voldman.
© 2012 John Wiley & Sons, Ltd. Published 2012 by John Wiley & Sons, Ltd.

1.2.1 Thales of Miletus and Electrostatic Attraction

Thales of Miletus, born in 624 B.C.E and died in 546 B.C.E, was the founder of the Ionian School (or Milesian School) and one of the Seven Wise Men of Ancient Greece in the Pre-Socratic era. Thales was an astronomer, mathematician, and philosopher. He was an inventor and an engineer. Thales of Miletus established a heritage of searching for knowledge for knowledge sake, development of the scientific method, establishment of practical methods, and the conjecture approach to questions of natural phenomenon. The Milesian School is regarded as establishing the critical method of questioning, debate, explanation, justification and criticism. The students of Thales included Euclid, Pythagoras, and Eudemus [2].

It was Thales of Miletus who was accredited with the discovery of the electrostatic attraction created after the material amber was rubbed. Thales noted after amber was rubbed, straw was attracted to the piece of amber. It was from this event, the Greek word for amber, ελεκτρον (trans. *electron*) became associated with the electrical phenomenon.

Knowledge of Thales' ideas was common through writings of his disciples and notary Greek Philosophers. In *De Anima 411 a7-8*, Aristotle stated "Some think that the soul pervades the whole universe, when perhaps came Thales' view that everything is full of gods." [3].

Electrostatic phenomenon pre-dated early thoughts of the nature of physical matter. Thales of Miletus's ESD experiments and study of electrostatic attraction was before the atomic schools of matter in Greece and Rome. Electrostatic phenomena and thought began before the Greek atomistic schools of Democritus (420 B.C.E.), and Epicurus (370 B. C.E.), and Roman School of Lucretius (50 B.C.E) Thales was deceased by the time the schools of atomic thought were active. On his tomb read, "Here in a narrow tomb Great Thales lie; Yet his renown for wisdom reached the skies [4]."

Robert A. Millikan, in the Introduction of his 1917 edition of "The Electron" [5] stated,

"Perhaps it is merely a coincidence that the man who first noticed rubbing of amber would induce in it a new and remarkable state now known as the state of electrification was also the man who first gave expression to the conviction that there must be some great unifying principle which links together all phenomena and is capable of making them rationally intelligible; that behind all the apparent variety and change of things there is some primordial element, out of which all things are made and the search for which must be the ultimate aim of all natural science. Yet if this be merely coincidence, at any rate to Thales of Miletus must belong a double honor. For he first correctly conceived and correctly stated, as far back as 600 B.C., the spirit which has actually guided the development of physics in all ages, and he also first described, though in a crude and imperfect way, the very phenomenon the study of which has already linked together several of the erstwhile isolated departments of physics, such as radiant heat, light, magnetism, and electricity, and has very recently brought us nearer to the primordial element than we have ever been before."

J.H. Jeans, in the 1925 Fifth Edition of *The Mathematical Theory of Electricity and Magnetism* [6], wrote

> *"The fact that a piece of amber, on being rubbed, attracted to itself other small bodies, was known to the Greeks, the discovery of this fact being attributed to Thales of Miletus"*
>
> *"A second fact, namely, that a certain mineral ore (lodestone) possessed the property of attracting iron, is mentioned by Lucretius. These two facts have formed the basis from which the modern science of Electromagnetism has grown."*

1.2.2 Electrostatics and the Triboelectric Series

With the death of Thales of Miletus, little progress proceeded on ESD phenomenon. Although history moved forward, the advancement of tribo-electric charging and electrostatic discharge phenomenon was slow relative to the time that passed. In Europe, mankind saw the Roman Empire, the Golden Age of Islam, the Middle Ages, the Black Death, the Renaissance, the Reformation, and the advancement of nation states. ESD phenomenon was discovered by Thales while China was undergoing the Zhou Dynasty. Asia underwent tremendous change with the Qin, Han, Sui, Tang, and Song, Yuan, and Ming dynastic periods, but there was no advancement of this field of knowledge.

With all this social change, insignificant growth in the understanding of electrostatic phenomenon increased until the eighteenth century. The interest in tribo-charging and electrostatic phenomenon became the luxury of scientists supported by the courts of Europe and laboratories of France and England.

So, how does tribo-charging happen?

When two materials come into contact, the atoms contained within the materials come into close contact. Figure 1.1 shows an example of two atoms of different materials. The nucleus is tightly bound, with neutrons and protons, and is positively charged. Strong interactions hold the neutrons and protons together. According to the Bohr model, the electrons orbit around the nucleus, and are drawn to the nucleus by the electrostatic attraction between the negative electrons and the positive nucleus. In a neutral un-charged atom, the number of protons and electrons are equal in number.

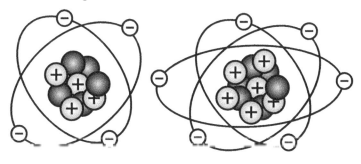

Material A   Material B

Figure 1.1   Tribo-charging of materials – physical contact

Negative Charged Atom   Positive Charged Atom

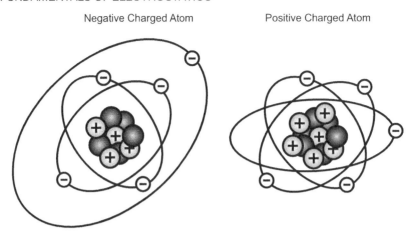

Figure 1.2   Tribo-charging of materials – separation

When two materials are in contact, charge transfer occurs through the friction or physical contact. An electron of the outer orbitals can transfer from one material to the second material. Figure 1.2 shows an example after the two materials are separated. In this case, the material that loses the electron becomes positively charged, and the material that gains the electron becomes negatively charged.

1.2.3 Triboelectric Series and Gilbert

Gilbert, in the seventeenth century, noted that interaction between a glass rod and silk produced the same phenomenon discussed by Thales of Miletus – materials when rubbed with silk became *"amberized"* [5]. Gilbert began construction of the earliest list of tribo-electrification.

1.2.4 Triboelectric Series and Gray

In the same period, Stephen Gray (1696–1736) thought of the concept of the division of materials according to its nature of removing or sustaining electrification [5]. He defined a class of materials which remove "amberization" as conductors, and the class of materials which allowed a body to retain its electrification as non-conductors, or insulators.

1.2.5 Triboelectric Series and Dufay

This work was followed by French physicist Dufay in 1733 discovering the same effect can be achieved with sealing wax and cat's fur and noted the effect was different from the glass rod [5]. Dufay first noted that there was an attractive and repulsive phenomenon between different materials, naming the opposite processes as "vitreous" and "resinous."

1.2.6 Triboelectric Series and Franklin

Benjamin Franklin – the first American ESD engineer, in 1747, also identified two processes, which he divided into "positive" and "negative" processes. The "positive" process was the first process, discovered by Gilbert, any physical body was electrified positive if repelled by a glass rod which was rubbed with silk. The "negative" process is any body repelled from sealing wax which was rubbed with cat's fur, extending the work of Dufay.

In this time frame, many electrostatic scientists began recording the relationship of one body to another in the electrification process. Lists of materials of the times were ordered to construct the early tribo-electrification chart [6]

> . . . "It is therefore possible to arrange any number of substances in a list such that a substance is charged with positive or negative electricity when rubbed with a second substance, according as the first substance stands above or below the second substance on the list. The following is a list of this kind which includes some of the most important substances:
> "Cat's skin, Glass, Ivory, Silk, Rock Crystal, The Hand, Wood, Sulphur, Flannel, Cotton, Shellac, Caouthouc, Resins, Guttapercha, Metals, Guncotton.""

J.H. Jeans D.Sc, LL.D, F.R.S
The Mathematical Theory of Electricity and Magnetism Cambridge, England, 1925

1.2.7 Electrostatics – Symmer and the Human Body Model

In this time frame, some electrostatic scientist and engineers explored phenomenon around them, and were the early root of electrostatic discharge (ESD) phenomenon of garments, and possibly early influences on the "human body model."

Robert Symmer (1759) would do experimental studies in the dark, exploring the electrical discharge phenomenon of removal of his stockings and the interactions of two different sets of stockings, placing two white stockings in one hand and two black stockings in the other [5].

1.2.8 Electrostatics – Coulomb and Cavendish

Coulomb (while other ESD scientists were still enjoying the ESD phenomenon with their stockings and rubbing things together) developed the Torsion Balance, in 1785, beginning a series of processes to understand the relationship of force, charge and physical distance. The experiments of Coulomb were also performed by Cavendish at an earlier date but not published until 1879 by James Clerk Maxwell [7].

1.2.9 Electrostatics – Faraday and the Ice Pail Experiment

At this time, the relationship of positive and negative electrification was not fully understood. In 1837, Michael Faraday performed the "ice-pail experiment" involving glass rod and silk which showed that "positive and negative electrical charges always appear simultaneously and in exactly equal amounts."

Faraday, in his lecture *Forces of Matter – Lecture V Magnetism-Electricity* would provide demonstrations of electrostatic charging phenomenon demonstrating the phenomenon and the relationship of positive and negative charging effects [8]. He would close his lecture on electrical phenomenon with ". . . *This, then is sufficient, in the outset to give you an idea of the nature of the force which we call ELECTRICITY. There is no end to the things from which you can evolve this power.*"

1.2.10 Electrostatics – Faraday and Maxwell

Even at this time frame, the relationship of matter, electrical charging and electrical force was not well understood. Different models were proposed from single electrical fluid models, and two fluid models to explain the charging process. Electrical phenomenon models were established to explain this phenomenon as related to a stress or strain of the medium, moving the thoughts from an atomistic perspective to a field representation. It was in this time frame that Faraday and James Clerk Maxwell began addressing the understanding of electricity and electrical forces in terms of a field perspective, and electrical charge was viewed as a "state of strain in the ether." James Clerk Maxwell, *in Electricity and Magnetism* in 1873, created the modern formulation of electricity and magnetism as we understand it today [7].

1.2.11 Electrostatics – Paschen

In 1889, Paschen began an analysis of breakdown phenomenon in gases trying to explain the relationship between gas pressure and electrode spacing [8]. Breakdown phenomenon in media took a great leap forward influencing electrostatic discharge (ESD) understanding in today's devices. Even today, the Paschen breakdown curve is important for understanding the electrical breakdown in air gaps and nano-structures.

1.2.12 Elcotrostatics – Stoney and the "Electron"

In 1891, the word "electron" as a natural unit of electricity was suggested by Dr. G. Johnstone Stoney, connecting the ideas of Faraday's Law of Electrolysis [5]. At this time, the understanding of the connection to physical matter was not understood, but was used as a measure or unit. Dr. G. Johnstone Stoney reconnected the idea to the Greek word for amber, connecting the early work of Thales of Miletus to the modern day concept of an electrical unit, later to be proven to be connected to matter and atomic theory.

From this brief history, it can be seen that electrostatic attraction pre-dates the earliest thoughts of the understanding of matter. From 600 B.C.E to even as late as the 1890s the relationship of the discovery of Thales of Miletus was not understood as connected to the transfer of electrons which were made of matter.

1.3 TRIBOELECTRIC CHARGING – HOW DOES IT HAPPEN?

Tribo-electric charging occurs when two bodies of materials come into physical contact, followed by separation of the two bodies. (Table 1.1) As discussed before, the word electron comes from the Greek word for "amber." The word "tribo" also comes from Greek, meaning "to rub." Tribo-charging has to do with two items that come into contact, followed by their separation. In the process of tribo-charging, electrons transfer from one material to another. Atoms are made of a nucleus which consists of positively charged protons, and neutrally charged neutrons. Outside of the nucleus, electrons establish standing wave orbitals around the nucleus. Electrons are negatively charged. When there are two dissimilar materials of different electric potentials, electrons can transfer from one material to another. After material separation, the material with the extra electron becomes negatively charged, whereas the material that lost the electron becomes positively charged.

So, the material property influences whether the electrons transfer. As discussed by Jeans [6], a chart can be constructed predicting the direction of charge transfer.

"It is therefore possible to arrange any number of substances in a list such that a substance is charged with positive or negative electricity when rubbed with a second substance, according as the first substance stands above or below the second substance on the list. . . "

Table 1.1 Triboelectric series

Positive (+)	Materials
	Rabbit Fur
	Glass
	Human Hair
	Nylon
	Wool
	Fur
	Silk
	Aluminum
	Paper
	Cotton
	Steel
	Wood
	Amber
	Nickel
	Gold
	Polyester
	Silicon
	Teflon
Negative (−)	

1.4 CONDUCTORS, SEMICONDUCTORS, AND INSULATORS

The electrical engineer likes to draw some defining boundary lines between insulators, semiconductors, and metals on the basis of such conductivity,

$$Insulators \ \sigma < 10^{-9} \ ohms^{-1} \ cm^{-1}$$

$$Semiconductor \ 10^{-9} < \sigma < 10^2 \ ohms^{-1} \ cm^{-1}$$

$$Metals \ \sigma > 10^2 \ ohms^{-1} \ cm^{-1}$$

Arthur Von Hippel "Building from Atoms" Chapter 2
The Molecular Designing of Materials and Devices, MIT Press, 1965

The physicist Professor Arthur Von Hippel simplifies the matter of what is an insulator, semiconductor, and metal based on the conductivity of a solid [9]. In layman's terms, an insulator is a material that limits the flow of electrons within the bulk volume or surface of the material, and a conductor is a material that allows free flow of electrons within the bulk or surface of the material. An insulator can have a surface resistance of $\rho_s = 10^{11}$ ohms and bulk resistance of $\rho = 10^{11}$ ohms-cm; whereas conductors can have a surface resistance less than $\rho_s = 10^4$ ohms and bulk resistance less than $\rho = 10^4$ ohms-cm. The boundaries between them are not well defined.

From a practical perspective, an insulator does not allow the free flow of electrons on its surface, or through its bulk, where in the case of semiconductors and metals, carriers can flow on and within the materials.

In semiconductor devices, components, and systems, the material properties can be a good thing, or a problem. In the case where a material does not allow the free flow of carriers, charge will build up, and can lead to the establishment of electric fields. High electric fields can lead to electrical breakdown, and electrical overstress. In the case where the material allows the free flow of carriers, high currents can establish. High currents can lead to high discharge currents, electrostatic discharge events, self-heating, thermal breakdown, and melting of components, packaging and systems.

1.5 STATIC DISSIPATIVE MATERIALS

In the field of electrostatic discharge protection, it is advantageous to have materials that are not highly insulating, and also not highly conducting. These materials are referred to as static dissipative materials. Static dissipative materials are the materials which are between the insulators and conductors. Hence, we can define them as materials in the range of $\rho_s = 10^4$ ohms to 10^{11} ohms surface resistance, and the range of $\rho = 10^4$ and 10^{11} ohms-cm bulk resistance. The advantage of having materials that are neither insulators or conductors is the avoidance of static charge buildup or high currents.

1.6 ESD AND MATERIALS

Unfortunately, components and systems are not that simple. Both electrical components and systems contain insulators, semiconductors, and metals. Semiconductor components comprise of insulating dielectrics, semiconductors, and metals.

Insulating dielectrics exist as inter-level dielectrics, thin film oxides, and buried oxide films in the substrate. Insulator films in the MOSFET gate dielectric are very thin films whose thickness is dependent on the technology generation. Inter-level dielectric films exist between the wiring levels in a semiconductor chip whose thickness is in the order of the metal wiring film thickness. Electrical breakdown of the dielectrics is a failure mechanism from electrical overstress (EOS), or electrostatic discharge (ESD) events. Insulators are used in systems for cards, boards, supports, and other components.

Metal films exist in semiconductor components as interconnects between the electronic circuits. Semiconductor components contain wire films consisting of aluminum and copper. Refractory metals, such as titanium, cobalt, tantalum, and tungsten are also used for contacts, vias, silicide films, to cladding for the wiring. In systems, metals are used for the system chassis, and shielding.

Semiconductor materials are used throughout semiconductor components as the base wafer for semiconductor devices. The conductivity varies dramatically through semiconductor devices using dopants, from resistive to conductive regions.

As a result, semiconductor components are a stratified medium of layers that consists of metals, insulators, and static dissipative materials. Hence, controlling the conduction and dissipative nature of a semiconductor component can be quite difficult with regions of high conducting and high insulating properties.

1.7 ELECTRIFICATION AND COULOMB'S LAW

"The force between two small charged bodies is proportional to the product of their charges, and is inversely proportional to the square of their distance apart, the force being one of repulsion or attraction according as the two charges are of the same or of opposite kinds."

Coulomb, 1785

From Coulomb's law, it is understood that the force between two objects is associated with the product of the charges, and inversely proportional to the square of the distance. Hence the electric field associated with a force on a "test charge" is the charge of the object, and inversely proportional to the square of the distance between.

1.7.1 Electrification by Friction

"Bodies may be electrified many other ways, as well as by friction"

> Part I Electrostatics Chapter I: *Electrification by Friction*
> *A Treatise on Electricity and Magnetism*
> James Clerk Maxwell, 1891

Electrification can be caused by rubbing two objects together. It is through friction between the surfaces (e.g., the materials touching) that can lead to the charge transfer and charging. When two materials are rubbed together, friction is assisting in the charge transfer process [7,10].

1.7.2 Electrification by Induction

"No force, either of attraction or of repulsion, can be observed between an electrified body and a body not electrified. When, in any case, bodies not previously electrified are observed to be acted on by an electrified body, it is because they have become electrified by induction"

> Part I Electrostatics Chapter I: *Electrification by Induction*
> *A Treatise on Electricity and Magnetism*
> James Clerk Maxwell, 1891

From an ESD perspective, charged materials and charged surfaces generate electric fields. These electric field lines start at the positive charge, and end at the negative charge.

Hence a system or component placed near a charged surface can be polarized where negative charges are formed on the surface. In this process of "induction", given that the object is electrically connected to a ground potential, polarization will occur. When the ground connection is removed, this object will be charged. Charge will flow either to or from the ground connection. In the "induction process", given that the ground connection and the electric field is removed, the component or system will remain charged [7,11].

Induction charging is a concern in semiconductor components which have been charged by external fields, and is not electrically grounded. As will be discussed, this is a concern for the "charged device model." When it is charged there is no concern; but, when the charge is rapidly discharged to a ground connection, high currents can lead to component damage.

1.7.3 Electrification by Conduction

James Clerk Maxwell noted that electrification can occur by placing a "wire" between a first object and a second, where the current flows from the first object to the second. In that time, Maxwell referred to this as "Electrification by Conduction [7]." Today, this would be regarded as charging a body through a voltage source and an electrical circuit. Hence, there are many ways to "charge" a body.

Today, in the manufacturing environment, electrical components, and systems, these charging processes occur and are part of the "events" that must be monitored, controlled, and eliminated.

1.8 ELECTROMAGNETISM AND ELECTRODYNAMICS

Electromagnetism has three branches, namely Electrostatics, Magnetostatics, and Electrodynamics. Electrostatics and Magnetostatics are independent of each other and address only states of rest. On the other hand, electrodynamics deals with the motion of electricity and magnetism [6].

1.9 ELECTRICAL BREAKDOWN

Although we regard electrostatic discharge (ESD) as an electrostatic phenomena, since charge is moving, and currents are flowing, it is electro quasi-static. Electrostatic discharge can be initiated by an arc from a person to an object. The electrostatic discharge event occurs as a result of the breakdown of the air in the gap between the person and the object. The discharge process is a function of spacing of the gap, the geometry of the gap (e.g., curvature, radius of electrodes), cleanliness of the surface, relative humidity, and speed of approach.

So, what is electrical breakdown?

1.9.1 Electrostatic Discharge and Breakdown

As a young graduate student at the MIT High Voltage Research Lab, my professor Markus Zahn had to do some research at Exxon in Linden New Jersey; so we were brought along to do some experimental work on the breakdown of oil . . . the researcher had a lab with a 200,000 Volt charging source, a large Corona ring, and a metal pipe to connect to the sample of oil. The breakdown of oil is a function of the oil purity, dirt, and oil degradation from discharge events. We were there to isolate the electrical instability after the oil breakdown occurrence and capture the signal during the event. We used an electro-optical isolating Kerr cell – nitrobenzene, two polarizers and a laser. The researcher worked six months to find this electrical oscillation of oil after breakdown . . . we were given 5 days – we found it after 4 days of exploration for the signal . . . Understanding the fundamentals of breakdown is important in gases, liquids and solids! . . . So this is what you do at MIT on your summer breaks!

Electrostatic discharge involves the breakdown in gases, liquids and solids [12]. Breakdown in gases, liquids or solids can be initiated by a feedback induced by the acceleration of carriers leading to secondary carriers. Breakdown phenomenon in air is important for ESD applications for spark gaps, ESD simulators, and magnetic recording industry. Today, there is a focus on the understanding of breakdown phenomenon for charged device simulators.

In the magnetic recording industry, breakdown can occur between the magneto-resistor (MR) element and the shields across the air bearing surface. Hence the physics of breakdown in air is relevant to today's problems. At very high speeds, the ability to provide semi-conductor devices may be limited. As a result, field emission devices and spark gaps may play a role in air bridge applications and micro-machines.

1.9.2 Breakdown and Paschen's Law

Paschen, in 1889, studied the breakdown physics of gases in planar gap regions [13]. The result of Paschen showed that breakdown process is a function of the product of the gas pressure and the distance between the electrodes. Paschen showed that

$$pd \approx \frac{d}{l}$$

where p is the pressure, d is the distance between the plates and l is the mean free path of the electrons. From the work of Paschen, a universal curve was established which followed the same characteristics independent on the gas in the gap. The Paschen curve is a plot of the logarithm of the breakdown voltage, V_{BD}, as a function of the logarithm of the product of the pressure and gap distance,

$$V_{BD} = f(pd)$$

At very low values of the p-d product, electrons must accelerate beyond the ionization limit to produce an avalanche process because the likelihood of impacts is too few. In this region, the breakdown voltage decreases with increasing value of the pressure-gap product. This occurs until a minimum condition is reached. At very high values of the pressure-gap product, the number of inelastic collisions is higher and the breakdown voltage increases. This U-shaped dependence is characteristic of gas phenomenon. At the high gas pressure, secondary processes, such as light emissions occur. Figure 1.3 shows the Paschen curve high-lighting its U-shaped characteristic.

1.9.3 Breakdown and Townsend

Avalanche phenomenon is important to understand the breakdown process in semiconductors and other materials. Townsend, in 1915, noted that the breakdown occurs at a critical ava-lanche height [12,14],

$$H = e^{\alpha d} = \frac{1}{\gamma}$$

In this expression, the avalanche height H, is equal to the exponential of the product of the probability coefficient of ionization (number of ionizing impacts per electron and unit

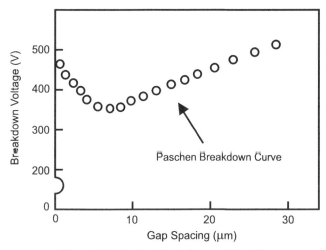

Figure 1.3 Paschen breakdown relationship

distance in the direction of the electric field) and electrode spacing. The avalanche height, H can also be expressed as the inverse of the probability coefficient of regeneration (number of new electrons released from the cathode per positive ion).

1.9.4 Breakdown and Toepler's Law

Evaluation of the resistance of arc discharges are important in ESD phenomenon since these events are evident in ESD simulation, such as charged device model (CDM), machine model (MM) for components and other ESD gun simulators for systems.

   Toepler, in 1906, established a relationship of the arc resistance in a discharge process [15]. Toepler's law states that the arc resistance at any time is inversely proportional to the charge which has flowed through the arc

$$R(t) = \frac{kTD}{\int\limits_0^t I(t')dt'}$$

where $I(t)$ is the current in the arc discharge at time t, and D is the gap between the electrodes. The value k_T is a constant whose value is 4×10^{-5} V-sec/cm.

1.9.5 Avalanche Breakdown

Avalanche breakdown plays an important role in the understanding of electrostatic discharge issues. Avalanche breakdown can lead to physical failure of semiconductor components.

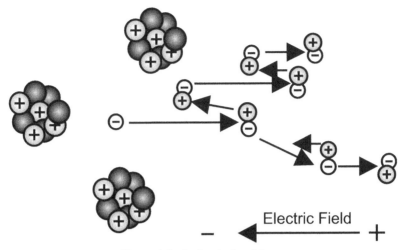

Figure 1.4 Avalanche breakdown

Avalanche breakdown also is used intentionally in semiconductors to serve as a trigger element to initiate turn-on of an ESD circuit. Avalanche breakdown plays a key role in semiconductor ESD elements and circuits which are intentionally reverse biased during functional operation of semiconductor chips, but are initiated at a voltage above the native operational voltage. This phenomenon is key to ESD networks in MOSFET and bipolar devices. It is utilized on input pin circuits, between common power rails in ESD power clamps, and between power rails. Hence, it is fundamental to the ESD discipline.

What is the avalanche process? Avalanche breakdown occurs when a carrier is accelerated by an electric field, where the collision with the material leads to a secondary carrier (Figure 1.4). The secondary carrier is then also accelerated, forming a third interaction. This is why it is called "avalanche multiplication."

To be more specific, it can involve energy transfer between the carriers, and the medium. As carriers are accelerated in a medium, energy is transferred from the electric field to the carriers. As the electric field increases, carriers approach a limiting drift velocity and further increases lead to thermal vibrations. As carriers are accelerated, there is a competition between energy transmitted to the electron and energy transmitted to the lattice. This is a function of the ionization threshold, the energy of the carrier, and mean free path of optical phonon scattering. To follow the cascade of carriers, the ionization rate is a function of the summation of probable events of cascades. The probability that a carrier contributes to ionization is a function of a first event where the carrier reaches the ionization and the probability that the first event is ionization. A second term is the probability that the carrier achieves the energy of a phonon and ionization event, and that the first event is phonon emission, followed by an ionization event. This is followed by a third term of the probability that an electron achieves the energy of two phonons and an ionization energy, where the probability is that two phonons are emitted followed by an ionization event. And on, and on. . . . From this collision process, we can derive an ionization coefficient associated with the avalanche collision processes.

Analysis of the breakdown in air is valuable for ESD protection to understand ESD simulators, and specifications. For example, D. Lin and T. Welsher related the phenomenon of air discharges to understand the physics of the first charged device model (CDM) test system [16].

Assuming a simple geometry of a gap with a gap spacing d, charge Q, and voltage V, we can evaluate the voltage, electric field, peak current, rise and fall time using the Townsend avalanche relationship. Assume a breakdown voltage and breakdown electric field is achieved in the gap leading to the flow of current across the gap structure. Let the current be a function of the electron drift current term (neglecting diffusion current).

$$I = en_e v_d$$

and

$$v_d = \mu E$$

We can relate the current to the derivative of the charge Q as a function of time, letting $dQ = (env)\, dt$, and we can relate the drift voltage to the electric field, and we know that the electric field, E, is a function of the voltage and gap spacing $E = V/d = Q/Cd$.

Assuming a breakdown voltage, and breakdown electric field magnitude, as the initial condition to initiate the discharge, we can derive the equations to explain the phenomena. Voltage, electric field, charge and current can be obtained in the gap when the electron density as a function of time is evaluated. Hence, all the terminal information can be expressed as a function of the charge density in time. In this fashion, the current as a function of time can be evaluated from the electric field, carrier density, and impact ionization coefficients. The peak current is a function of the capacitance, the gap size, the mobility and the ionization at the maximum electric field.

Air breakdown is important in ESD phenomenon and the physics of failure. A key parameter of interest is the peak current of the air discharge since a number of ESD failure mechanisms is associated with the peak current magnitude. It can be shown that the peak maximum current is expressable as [16]

$$I_{Peak} = (\mu Cd)E_{max}^2 \alpha(E_{max})$$

The peak current is then a function of the capacitance, the gap size, the mobility and the ionization at the maximum electric field.

1.10 ELECTROQUASISTATICS AND MAGNETOQUASISTATICS

For static electric and magnetic fields, it is adequate to address evaluation in the electro quasistatic and magnetoquastatic approximation of Maxwell's equations. The approximation is a function of the charge relaxation time, magnetic diffusion time, and the time constant of interest.

1.11 ELECTRODYNAMICS AND MAXWELL'S EQUATIONS

James Clerk Maxwell showed that there exists a set of equations that explain the connection within electromagnetism between electric and magnetic fields [7]. Electric charge leads to static electric fields, and current leads to magnetic fields. Additionally, time varying fields lead to sources that influence electrical components, and systems.

In production and manufacturing, one must not only be concerned with static fields, but time varying fields that produce electromagnetic interference (EMI), and electromagnetic compatibility (EMC) issues.

1.12 ELECTROSTATIC DISCHARGE (ESD)

Today, the discipline of electrostatic discharge (ESD) is focused on the impact to components and systems. Prior to the 1990s, there were few texts associated with the field of electrostatic discharge. In the last twenty years, many texts have been written discussing the ESD design discipline and latchup design discipline [17–30].

As will be discussed further in the text, there are many new ESD test standards to qualify and evaluate both components and systems. Components tests consist of tests for human body model (HBM), machine model (MM), charged device model (CDM), transmission line pulse testing (TLP), and very fast transmission line pulse (VF-TLP) testing [31–43].

For system level evaluation, there are tests for cable discharge events [44–46], system test IEC 61000–2, and human metal model (HMM) [47–53].

1.13 ELECTROMAGNETIC COMPATIBILITY (EMC)

Electromagnetic compatibility (EMC) is the ability of an electronic system to function properly in its intended electromagnetic environment and not be a source of electronic emissions to that electromagnetic environment. Electromagnetic compatibility (EMC) has two features. A first feature is a source of emission of an electromagnetic field. A second feature is the collector of electromagnetic energy. The first aspect is the emission of an electromagnetic field which may lead to electromagnetic interference of other components or systems. The second aspect has to do with susceptibility of a component, or system to the undesired electromagnetic field. Today, there are many standards and tests on the subject of EMC [54–75].

1.14 ELECTROMAGNETIC INTERFERENCE (EMI)

Electromagnetic interference (EMI) is interference, or noise, generated from an electromagnetic field. Electromagnetic interference is electric and magnetic fields that interfere with electrical components, magnetic components, and electrical or magnetic systems. EMI can lead to both component level or system level failure of electronic systems. EMI can lead to failure of electronic components, without physical contact to the electronic system. In the industry, there is a significant number of standards and tests to address both EMC and EMI concerns [54–76].

1.15 SUMMARY AND CLOSING COMMENTS

In Chapter 1, a brief discussion of electrostatics, and triboelectrical phenomena was weaved into the individuals, dates, and history: Thales of Miletus, Gray, Dufay, Franklin, Toepler, Faraday, Cavendish, Coulomb, to Maxwell – just to mention a few. The chapter then quickly fast-forwards to today's issues of ESD, EMC and EMI, all today's concerns in components and systems.

In Chapter 2, the text discusses the manufacturing and factory environment, and how it is related to the discussion in Chapter 1. Today, the factory environment is established to address everything we learned in Chapter 1 about materials, conductivity and avoidance of discharge events. What will be observed is how it has been integrated and applied into the manufacturing environment and business process to reduce the impact of ESD phenomena.

. . . Thales of Miletus would be shocked to see how far this has gone!

REFERENCES

1. Voldman, S. (2002) Lightning rods for nanoelectronics. *Scientific American*, vol, 287, no. 4, 90–97.
2. (1944) *Philologus*, **96**, 170–182.
3. Aristotle. *De Anima*, **411**, a7–a8.
4. Kirk, G.S., Raven, J.E., and Schofield., M. (1995) *The Pre-Socratic Philosophers*, 2nd edn, Cambridge University Press.
5. Millikan, R.A. (1917) *The Electron*, University of Chicago Press.
6. Jeans, J.H. (1925) *The Mathematical Theory of Electricity and Magnetism*, Fifth edn, Cambridge University Press.
7. Maxwell, J.C. (1873) A Treatise on Electricity and Magnetism.
8. Faraday, M. (1910) *The Forces of Matter, Lecture V: Electricity, Scientific Papers, Harvard Classics*, vol. **30**, P.F. Colliers & Sons Company, New York, pp. 62–74.
9. Von Hippel, A. (1965) Building from atoms, Chapter 2, in *The Molecular Designing of Materials and Devices*, MIT Press, Cambridge, Massachusetts, pp. 9–28.
10. Thomson, Sir W. (March 1848) *On the Mathematical Theory of Electricity in Equilibrium*, Cambridge and Dublin Mathematical Journal, Cambridge, England.
11. Faraday, M. (1843) On static electrical induction action. *Philosophy Magazine*.
12. Paschen, F. (1889) Ueber die zum Funkenübergang in Luft, Wasserstoff und Kohlensäure bei verschiedenen Drucken erforderliche Potentialdifferenz, *Annals of Physics*, vol. 273, no. 5, 69–86.
13. Von Hippel, A. (1965) Conduction and breakdown, in *The Molecular Designing of Materials and Devices*, MIT Press, Cambridge, Massachusetts, pp. 183–197.
14. Townsend, J.S. (1915) *Electricity in Gases*, Clarendon Press, Oxford.
15. Toepler, M. (1906) Uber Funkenspannungen *Annalen der Physik*, vol. 324, no. 1, **191–209**, 191.
16. Lin, D. and Welsher, T. (1992) From lightning to charged device model electrostatic discharges. Proceedings of the Electrical Overstress/Electrostatic Discharge (EOS/ESD) Symposium, pp. 68–75.

17. Dabral, S. and Maloney, T.J. (1998) *Basic ESD and I/O Design*, John Wiley and Sons Ltd., West Sussex.
18. Wang, A.Z.H. (2002) *On Chip ESD Protection for Integrated Circuits*, Kluwer Publications, New York.
19. Amerasekera, A. and Duvvury., C. (2002) *ESD in Silicon Integrated Circuits*, 2nd edn, John Wiley and Sons, Ltd., West Sussex.
20. Gossner, H., Esmark, K., and Stadler, W. (2003) *Advanced Simulation Methods for ESD Protection Development*, Elsevier Science Publication.
21. Voldman, S. (2004) *ESD: Physics and Devices*, John Wiley and Sons, Ltd., Chichester, England.
22. Voldman, S. (2005) *ESD: Circuits and Devices*, John Wiley and Sons, Ltd., Chichester, England.
23. Voldman, S. (2006) *ESD: RF Circuits and Technology*, John Wiley and Sons, Ltd., Chichester, England.
24. Voldman, S. (2007) *Latchup*, John Wiley and Sons, Ltd., Chichester, England.
25. Voldman, S. (2008) *ESD: Circuits and Devices*, Publishing House of Electronic Industry (PHEI), Beijing, China.
26. Voldman, S. (2009) *ESD: Failure Mechanisms and Models*, John Wiley and Sons, Ltd., Chichester, England.
27. Mardiquan, M. (2009) Electrostatic discharge, in *Understand, Simulate, and Fix ESD Problems*, John Wiley and Sons, Co., New York.
28. Ker, M.D. and Hsu, S.F. (2009) *Transient Induced Latchup in CMOS Integrated Circuits*, John Wiley and Sons, Ltd., Singapore.
29. Vashchenko, V. and Shibkov, A. (2010) *ESD Design in Analog Circuits*, Springer, New York.
30. Voldman, S. (2009) *ESD: Design and Synthesis*, John Wiley and Sons, Ltd., Chichester, England.
31. ANSI/ESD ESD-STM 5.1 – 2007 (2007) ESD Association Standard Test Method for the Protection of Electrostatic Discharge Sensitive Items - Electrostatic Discharge Sensitivity Testing - Human Body Model (HBM) Testing - Component Level. Standard Test Method (STM) document.
32. ANSI/ESD SP 5.1.2-2006 (2006) ESD Association Standard Practice for the Protection of Electrostatic Discharge Sensitive Items - Human Body Model (HBM) and Machine Model (MM) Alternative Test Method: Split Signal Pin-Component Level.
33. ANSI/ESD ESD-STM 5.2 – 1999 (1999) ESD Association Standard Test Method for the Protection of Electrostatic Discharge Sensitive Items - Electrostatic Discharge Sensitivity Testing - Machine Model (MM) Testing - Component Level. Standard Test Method (STM) document.
34. ANSI/ESD ESD-STM 5.3.1 – 1999 (1999) ESD Association Standard Test Method for the Protection of Electrostatic Discharge Sensitive Items - Electrostatic Discharge Sensitivity Testing – Charged Device Model (CDM) Testing - Component Level. Standard Test Method (STM) document.
35. Voldman, S., Ashton, R., Barth, J. *et al.* (2003) Standardization of the transmission line pulse (TLP) methodology for electrostatic discharge (ESD). Proceedings of the Electrical Overstress/Electrostatic Discharge (EOS/ESD) Symposium, pp. 372–381.
36. ANSI/ESD Association ESD-SP 5.5.1-2004 (2004) ESD Association Standard Practice for the Protection of Electrostatic Discharge Sensitive Items - Electrostatic Discharge Sensitivity Testing – Transmission Line Pulse (TLP) Testing Component Level. Standard Practice (SP) document.
37. ANSI/ESD Association ESD-STM 5.5.1-2008 (2008) ESD Association Standard Test Method for the Protection of Electrostatic Discharge Sensitive Items - Electrostatic Discharge Sensitivity Testing – Transmission Line Pulse (TLP) Testing Component Level. Standard Test Method (STM) document.
38. ANSI/ESD STM5.5.1-2008 (2008) Electrostatic Discharge Sensitivity Testing – Transmission Line Pulse (TLP) – Component Level.

39. ANSI/ESD STM5.5.2-2007 (2007) Electrostatic Discharge Sensitivity Testing - Very Fast Transmission Line Pulse (VF-TLP) - Component Level.

40. ESD Association ESD-SP 5.5.2 (2007) ESD Association Standard Practice for the Protection of Electrostatic Discharge Sensitive Items - Electrostatic Discharge Sensitivity Testing Very Fast Transmission Line Pulse (VF-TLP) Testing Component Level. Standard Practice (SP) document.

41. ANSI/ESD Association ESD-SP 5.5.2-2007 (2007) ESD Association Standard Practice for the Protection of Electrostatic Discharge Sensitive Items - Electrostatic Discharge Sensitivity Testing – Very Fast Transmission Line Pulse (VF-TLP) Testing Component Level. Standard Practice (SP) document.

42. ESD Association ESD-STM 5.5.2 (2009) ESD Association Standard Test Method for the Protection of Electrostatic Discharge Sensitive Items - Electrostatic Discharge Sensitivity Testing Very Fast Transmission Line Pulse (VF-TLP) Testing Component Level. Standard Test Method (STM) document.

43. ANSI/ESD Association ESD-STM 5.5.1-2008 (2008) ESD Association Standard Test Method for the Protection of Electrostatic Discharge Sensitive Items - Electrostatic Discharge Sensitivity Testing – Very Fast Transmission Line Pulse (VF-TLP) Testing Component Level. Standard Practice (SP) document.

44. ESD Association DSP 14.1-2003 (2003) ESD Association Standard Practice for the Protection of Electrostatic Discharge Sensitive Items – System Level Electrostatic Discharge Simulator Verification Standard Practice. Standard Practice (SP) document.

45. ESD Association DSP 14.3-2006 (2006) ESD Association Standard Practice for the Protection of Electrostatic Discharge Sensitive Items – System Level Cable Discharge Measurements Standard Practice. Standard Practice (SP) document.

46. ESD Association DSP 14.4-2007 (2007) ESD Association Standard Practice for the Protection of Electrostatic Discharge Sensitive Items – System Level Cable Discharge Test Standard Practice. Standard Practice (SP) document.

47. International Electro-technical Commission (IEC) IEC 61000-4-2 (2001) Electromagnetic Compatibility (EMC): Testing and Measurement Techniques – Electrostatic Discharge Immunity Test.

48. Grund, E., Muhonen, K., and Peachey, N. (2008) Delivering IEC 61000-4-2 current pulses through transmission lines at 100 and 330 ohm system impedances. Proceedings of the Electrical Overstress/Electrostatic Discharge (EOS/ESD) Symposium, pp. 132–141.

49. IEC 61000-4-2 (2008) Electromagnetic Compatibility (EMC) – Part 4-2:Testing and Measurement Techniques – Electrostatic Discharge Immunity Test.

50. Chundru, R., Pommerenke, D., Wang, K. *et al.* (2004) Characterization of human metal ESD reference discharge event and correlation of generator parameters to failure levels – Part I: Reference Event. *IEEE Transactions on Electromagnetic Compatibility*, **46** (4), 498–504.

51. Wang, K., Pommerenke, D., Chundru, R. *et al.* (2004) Characterization of human metal ESD reference discharge event and correlation of generator parameters to failure levels – Part II: Correlation of generator parameters to failure levels. *IEEE Transactions on Electromagnetic Compatibility*, **46** (4), 505–511.

52. ESD Association ESD-SP 5.6-2008 (2008) ESD Association Standard Practice for the Protection of Electrostatic Discharge Sensitive Items - Electrostatic Discharge Sensitivity Testing – Human Metal Model (HMM) Testing Component Level. Standard Practice (SP) document.

53. ANSI/ESD SP5.6-2009 (2009) Electrostatic Discharge Sensitivity Testing - Human Metal Model (HMM) - Component Level.

54. Jowett, C.E. (1976) *Electrostatics in the Electronic Environment*, Halsted Press, New York.

55. Lewis, W.H. (1995) *Handbook on Electromagnetic Compatibility*, Academic Press, New York.

56. Morrison, R. and Lewis, W.H. (1990) *Grounding and Shielding in Facilities*, John Wiley and Sons Inc., New York.

57. Paul, C.R. (2006) *Introduction to Electromagnetic Compatibility*, John Wiley and Sons Inc., New York.

58. Morrison, R. and Lewis, W.H. (2007) *Grounding and Shielding*, John Wiley and Sons Inc., New York.

59. Ott, H.W. (2009) *Electromagnetic Compatibility Engineering*, John Wiley and Sons Inc., Hoboken, New Jersey.

60. Ott, H.W. (1985) Controlling EMI by proper printed wiring board layout. Sixth Symposium on EMC, Zurich, Switzerland.

61. ANSI C63.4-1992 (July 17 1992) *Methods of Measurement of Radio-Noise Emissionss from Low-Voltage Electrical and Electronic Equipment in the Range of 9 kHz to 40 GHz*, IEEE.

62. EN 61000-3-2 (2006) *Electromagnetic Compatibility (EMC) – Part 3-2: Limits-Limits for Harmonic Current Emissions (Equipment Input Current < 16 A Per Phase)*, CENELEC.

63. EN 61000-3-3 (2006) *Electromagnetic Compatibility (EMC) – Part 3-3: Limits-Limitation of Voltage Changes, Voltage Fluctuations and Flicker in Public Low-Voltage Supply Systems for Equipment with Rated Current <16 A Per Phase and Not Subject to Conditional Connection*, CENELEC.

64. EN 61000-4-2 (2001) Electromagnetic Compatibility (EMC) – Part 4-2: Testing and Measurement Techniques – Electrostatic Discharge Immunity Test.

65. MDS MDS-201-0004 (October 1 1979) *Electromagnetic Compatibility Standards for Medical Devices*, U.S. Department of Health Education and Welfare, Food and Drug Administration.

66. MIL-STD-461E (August 20 1999) Requirements for the Control of Electromagnetic Interference Characteristics of Subsystems and Equipment.

67. RTCA RTCA/DO-160E (December 7 2004) *Environmental Conditions and Test Procedures for Airborne Equipment*, Radio Technical Commission for Aeronautics (RTCA).

68. SAE SAE J551 (June 1963) *Performance Levels and Methods of Measurement of Electromagnetic Compatibility of Vehicles and Devices (60 Hz to 18 GHz)*, Society of Automotive Engineers.

69. SAE SAE J1113 (June 1995) *Electromagnetic Compatibility Measurement Procedure for Vehicle Component (Except Aircraft) (60 Hz to 18 GHz)*, Society of Automotive Engineers.

70. Wall, A. (2004) Historical Perspective of the FCC Rules for Digital Devices and a Look to the Future. IEEE International Symposium on Electromagnetic Compatibility, August 9–13, 2004.

71. Denny, H.W. (1983) *Grounding for the Control of EMI*, Don White Consultants, Gainesville, VA.

72. Boxleitner, W. (1989) *Electrostatic Discharge and Electronic Equipment*, IEEE Press, New York.

73. Gerke, D. and Kimmel, W.D. (1994) The Designer's Guide to Electromagnetic Compatibility, EDN, vol. 39, no. 2, pp. S3-S114.

74. Kimmel, W.D. and Gerke, D.D. (1993) Three keys to ESD system design. *EMC Test and Design*.

75. Violette, J.L.N. and Violette, M.F. (1986) ESD case history – Immunizing a desktop business machine. *EMC Technology*, May–June 1986, vol. **4**, 55–60.

76. Wong, S.W. (1984) ESD design maturity test for a desktop digital system. *Evaluation Engineering*, vol. **23**, 104–112.

2 Fundamentals of Manufacturing and Electrostatics

A customer had a package with the company label on the semiconductor chip. The package had a metal top with the name of the company painted on the metal sur face. The manufacturing engineer did not want the customer to be upset from scratches on the label, so he put plastic tape covering the conveyer belt that the chips were on.

Two shifts worked on picking up the package and placing them on ESD safe foam and packaging so that the parts would not be impacted from shipping. Sixty six percent of the microprocessors were failing functionality after assembly in the packaging facility. Out of the parts, 200 out of 300 chips were failing. Experts on yield analysis studied lot-to-lot and wafer-to-wafer dependency. Yield experts evaluated which tools the parts went through during semiconductor manufacturing. It was discovered that the failures were shift related.

On the first shift, one operator in the packaging assembly plant was tall and slender; the second shift, the other operator was short. The tall slender operator did not touch the pins when she removed them from the conveyer belt. The other operator's fingers touched the pins.

Manufacturing environments are a complex arena which includes buildings, air handling, flooring, tooling, machines, equipment, operators, technicians, and electrical components. In these complex environments, it is difficult to provide all the proper controls to avoid electrostatic discharge (ESD), electrical overstress (EOS), static electric and magnetic fields, to electromagnetic interference (EMI).

ESD Basics: From Semiconductor Manufacturing to Product Use, First Edition. Steven H. Voldman.
© 2012 John Wiley & Sons, Ltd. Published 2012 by John Wiley & Sons, Ltd.

One of the key challenges in manufacturing is to provide an environment that does not impact the products being constructed in a manufacturing line, product assembly, or shipping and handling.

In the early development of electrostatic discharge phenomena, this was the primary challenge – how do you create a manufacturing environment free of ESD problems? How do you find ESD concerns? How do you qualify a manufacturing line? How do you monitor a manufacturing line? How do you check and verify that the manufacturing line is in compliance with specific requirements? Fundamentally, how do you run a business and be successful without ESD concerns from manufacturing to shipping?

In the late 1970s and 1980s, the issue of measuring, monitoring, evaluating, checking, and verifying ESD was a key issue. ESD awareness, ESD control programs, to ESD program management achieved a high focus [1–18]. Early leaders in this area of ESD program management were McAteer, McFarland, Halperin, and Dangelmayer. The early programs from the late 1970s, and 1980s, as well as other internal corporate ESD management programs, became the basis for today's ESD program management systems, such as S20.20 (e.g., for example, today's S20.20 ESD control program was an outgrowth of the IBM internal auditing and control system) [19–24].

In parallel with the ESD program management systems, technical reports, and ESD standards were developed to measure, and verify compliance of ESD protective equipment and materials [23]. Today, there are ESD standards for all materials, equipment, and tooling in the ESD Protected Area (EPA).

In this chapter, the subject of ESD control, measurement and verification is addressed for manufacturing environments. In this chapter, we will discuss test equipment, protective equipment, and materials. This discussion will be followed by ESD control programs and audits.

2.1 MATERIALS, TOOLING, HUMAN FACTORS, AND ELECTROSTATIC DISCHARGE

ESD concerns in manufacturing are a combination of the materials, tooling, and the human factors. The materials influence the triboelectric charge transfer. The tooling used can lead to charge transfer, and operators can participate in this transfer process.

In the manufacturing area, the electric field between the ceiling and the floor is influenced by the height of the ceiling, air flow, and placement of the ionizers. The placement of the ionizers relative to the worksurface where the sensitive parts are placed influence the effectiveness of the ionizers. The worksurface material, and its physical size is also a factor.

For tunneling magneto-resistors (TMR) in the magnetic recording industry, where human body model (HBM) levels are below 10 V HBM sensitivities, manufacturing sectors with small worksurfaces, and only a few operators at a given table. The air ionizers are placed very low near the worksurface and sensitive parts.

2.1.1 Materials and Human Induced Electric Fields

Operators in the manufacturing line can influence the triboelectric charging process. All external surfaces of the operator, the type of materials, and the proximity of the operator to the item can influence the charge transfer, and the human-induced electric field imposed. The footwear, the garments, the wrist straps, the personnel grounding of the garments can all influence the impact of the operator on tribocharging, and the ESD discharge. Additionally, the seating, position, and the distance of the operator from the sensitive parts can also influence the electric fields.

2.2 MANUFACTURING ENVIRONMENT AND TOOLING

In the transition from 200 to 300 mm wafers, the size of the manufacturing chambers, the electrostatic wafer chucks, and wafers were increased in size. The lateral electric field across the wafer became an issue in the plasma etch chambers. Plasma arcs occurred across the wafer, leading to destruction of a few sites on the wafer. At the location, the chip guard ring, vias and metal levels were melted, and significant damage was evident in the semiconductor chip (Figure 2.1).

2.3 MANUFACTURING EQUIPMENT AND ESD MANUFACTURING PROBLEMS

In wafer dicing operation, a saw is used to cut through the silicon wafer to separate the semiconductor chips. As the blade cuts through the saw, a liquid jet stream is

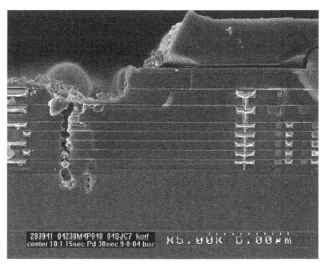

Figure 2.1 Plasma arcing

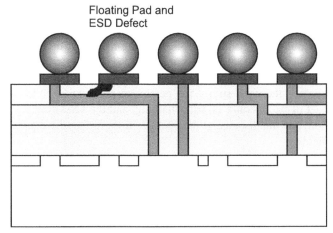

Figure 2.2 Floating solder balls

required in the dicing operation. In the stream, ions are present within the stream itself leading to buildup of charge on the semiconductor wafer.

Solder balls and bond pads on the wafer can collect the charge during the operation, leading to charging of the solder ball/bond pad structure. If the bond pad, or the solder ball/bond pad structure is not connected to circuitry, the structure charges until the inter-level dielectric (ILD) insulator breaks down, causing cracks in the insulator. During humidity testing, these insulator cracks can lead to moisture entering the semiconductor chip.

During ESD simulation testing, the "floating solder balls" were tested to the HBM test standard. In one technology, the ESD failures occurred at 1200 V HBM. In the scaled technology generation, the HBM ESD failures occurred at 900 V HBM (Figure 2.2).

2.4 MANUFACTURING MATERIALS

The choice of ESD materials in a manufacturing environment can have a large effect on the ESD protected area (EPA). The material choice can influence its initial conductivity, as well as the conductivity as a function of time. Material coatings and cleaning processes can influence the material conductivity. The wear-out of a floor or garment can influence its global conductivity as well as its spatial variation. It is these factors that show why it is important to qualify a manufacturing environment, establish a measurement set of procedures, and temporal audits of the items used in the manufacturing sector.

2.5 MEASUREMENT AND TEST EQUIPMENT

So, how does one compartmentalize the manufacturing environment? The manufacturing environment consists of the following categories (Figure 2.3) [23]:

Figure 2.3   ESD in manufacturing

- Grounding and Bonding Systems.
- Worksurfaces.
- Wrist Straps.
- Monitors.
- Footwear.
- Flooring.
- Personnel Grounding with Garments.
- Ionizers.
- Seating.
- Mobile Equipment.
- Packaging.

2.5.1 Manufacturing Testing for Compliance

Manufacturing test equipment is needed for evaluation of compliance to specifications. ESD test equipment includes the following (Figure 2.4):

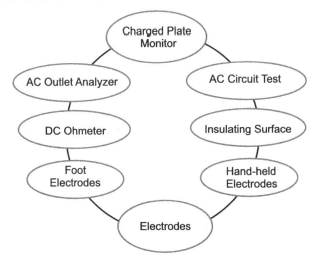

Figure 2.4 Manufacturing test equipment

- DC Ohmmeter.

- Electrodes.

- Handheld Electrodes.

- Foot Electrode.

- AC Outlet Analyzer.

- AC Circuit Tester (Impedance Meter).

- Insulative Support Surface.

- Charged Plate Monitor.

For all these items, it is necessary to verify the electrical measurements. In order to verify compliance, electrical measurements will be the means of determining an "ESD safe" environment, and compliance with objectives.

For evaluating the manufacturing environment, a dc ohmmeter needs to have a 10% accuracy (plus or minus) and be able to measure a direct current resistance in the range of 0.1 Ohm to 1.0 megaohms [23].

For electrodes, specifications are defined for electrodes, hand-held electrodes, and a foot electrode. For the standard electrode there should be a cylinder electrode weighing 5 lbs with a range of plus or minus 2 ounces. The diameter of the cylinder is defined as 2.5 inches with a plus or minus range of 0.1 inch. A contact area has a Shore-A (IRHD) durometer hardness between 50 and 70 [23]. For a handheld electrode, the dimensions are different; a hand-held stainless steel electrode is 3 inches or longer, and 1 inch in diameter, with a banana plug attached to one end. For the foot electrode, it is a 12 inch square conductive metal plate.

For an alternating current, a few items are needed. An alternating current (ac) outlet analyzer is needed to determine the presence of the equipment grounding conductor. An ac circuit tester is needed to measure impedance of the equipment grounding conductor.

In addition, to close the set of items to evaluate and verify compliance of a manufacturing line an insulator support surface whose resistance should be greater than 1.0×10^{13} ohms per square is required [23].

Lastly, a charged plate monitor (CPM) or portable verification kit is needed for evaluate and verify conformance. A portable verification kit has a field meter, isolated conductive plate, a plate charger, and an electrical ground cord.

2.6 GROUNDING AND BONDING SYSTEMS

For the manufacturing environment, it is very important to have an effective grounding and bonding system in place to avoid unwanted electrostatic discharge (ESD) failures. ESD test procedures, such as ANSI/ESD S6.1 Grounding, and the ANSI/NFPA 70 National Electric Code, provide guidance for proper grounding [26,27].

In the ESD protective workstation, a "Common Point Ground" is established for ESD protective worksurface, ESD protective mats, desk drawers, fixtures, personnel ground [23]. For verification of compliance to a "Common Point Ground," ac outlet analyzers, and ac circuit testers can be used. As a means to verify an ESD safe area, the connections between the personnel and the common point ground (personnel ground points), and the electrical receptacle must be checked for failures.

Equi-potential bonding is also important to avoid electrostatic discharge between two physical objects. Note that when all items in the area of an operator and sensitive parts are at the same electrical potential, no ESD event occurs.

2.7 WORKSURFACES

Worksurfaces can influence the electric field and charge deposition in a manufacturing environment. Since in ESD sensitive parts, whether single chips, trays, or system level cards or boards, the charged state of the worksurface is key in providing an ESD safe work area. Non-uniform conductivity in a worksurface can lead to electrical failures. Hence verification of the worksurface is critical for a safe ESD section.

ESD standards that apply to worksurfaces are ESD ADV53.1 ESD Protective Workstations, ANSI/ESD S4.1 Worksurfaces – Resistance Measurements, and the previous referenced ANSI/ESD 6.1 Grounding reference [32–34]. In these tests, the objective is to verify that the worksurface is electrically bonded to a grounded reference point. Test procedure factors that influence the test are as follows:

- Surface Cleaning.

- Device Under Test (DUT) Removal.

- Electrical Connections.

- Electrode Placement.

- Integrated Checker Stabilization.

2.8 WRIST STRAPS

Wrist straps are used in manufacturing environments to prevent the charge that is generated on personnel to be transferred to the electronic components, or systems.

Charge accumulates on personnel due to triboelectric charging between the flooring or other surfaces. The personnel stores the charge until contact occurs with other items. When the personnel touch electronic components, the charge is transferred along the arm and garment of the personnel causing an electrostatic discharge to the components, or tooling. With the use of a low resistance wrist strap, the current will flow through the wrist strap to the electrical ground of the workstation instead of the ESD sensitive parts.

To ensure the effectiveness of the wrist strap, the impedance to the electrical ground path must be lower than to the components. Hence, the objective of wrist strap integrity is to verify that the total resistance of all the elements in the wrist strap system is within an allowable resistance range. Standards and technical reports for wrist straps, such as ANSI/ESD S1.1-2006 Wrist Straps, and ESD TR1.0-01-01, Survey of Constant (Continuous) Monitors for Wrist Straps exist to quantify the specifications, and monitor wrist strap integrity and compliance [35,36].

2.9 CONSTANT MONITORS

Monitors are needed to verify integrity of different items in the manufacturing area, such as wrist straps. An example of a constant monitor is the constant monitor for wrist strap systems, ESD TR1.0-01-01, Survey of Constant (Continuous) Monitors for Wrist Straps [36]. Electrical connections between ground points, wrist strap, and the operator's body are tested during handling of components. These constant monitors over time reflect changes in the integrity of the manufacturing ESD safe area.

2.10 FOOTWEAR

Footwear, foot grounders, and interaction with the flooring are important to control and limit the tribocharging of manufacturing operators. Standards exist to characterize and measure the resistance of personnel, such as ANSI/ESD STM9.1-2006 Footwear – Resistive Characterization, ESD SP9.2-2003 Footwear – Foot Grounders Resistive Characterization, and ANSI/ESD STM97.1-2006 Floor Materials and Footwear – Resistance Measurements in Combination with A Person [37–39]. Footwear may consist of shoes, foot grounders, and manufacturing booties. Manufacturing environments constantly verify footware resistance using integrated checkers, meters, hand-held electrodes, and foot electrodes. During the compliance verification procedure, the resistance is measured to insure the resistance is within an acceptable range.

2.11 FLOORS

In a manufacturing environment, flooring influences the tribo-charging between the floor and personnel, and mobile equipment. The flooring charging characteristics are a function of the material property, installation, coatings, finish, paints and mats. Standard exist for both the footwear and the flooring such as ANSI/ESD STM97.1-2006 Floor Materials and Footwear – Resistance Measurements in Combination with A Person, and ANSI/ESD S7.1 – 2005 Resistive Characterization of Materials – Floor Materials [39,40]. The testing of the floor's resistance is achieved using an integrated checker, an electrode and an equipment ground point; this will provide a resistance point-to-ground evaluation.

2.12 PERSONNEL GROUNDING WITH GARMENTS

An important evaluation is the grounding of the personnel with garments. Personnel grounding with garments are an important test for the manufacturing environment. The procedure for verification and compliance is partly contained within the ANSI/ESD S1.1 Wrist Straps standard [41].

2.12.1 Garments

The garments used in the manufacturing environment influence the static charge accumulation on personnel as well as charge transfer during electrostatic discharge (ESD) events. A standard discussing the requirement and specification of garments is ANSI/ESD STM2.1-Garments [42].

2.13 AIR IONIZATION

Air ionization is used to neutralize static charge on insulator surfaces, worksurfaces, or isolated objects. Air ionizers provide a source of both positive and negative ions. Static charge will be neutralized by attraction of the ions of opposite polarity. Different type of air ionizers are used in the ionization process. Air ionizers are used in clean rooms, where air ionization may be one of the only allowable methods for electrostatic discharge control. The effectiveness of the air ionizers is a function of the applied voltage, the ionizer electrode tips, the ionizer tip cleanliness, and the height of the ionizer above the worksurface. For example, in the magnetic recording industry, with very sensitive tunneling magneto-resistors (TMR), ionizers are placed very low above the worksurface. Ionization standards include ANSI/ESD STM3.1 – 2006 Ionization, and ANSI/ESD SP3.3-2006 Periodic Verification of Air Ionizers [43,44]. Figure 2.5 shows a picture of an ionizer array probe tips and positive and negative ions. Figure 2.6 shows a commercial ionizer used in semiconductor manufacturing environments.

2.14 SEATING

When personnel sit in chairs in a manufacturing environment, it is ideal that the personnel are discharged to a ground potential. Seating in a manufacturing environment can be on

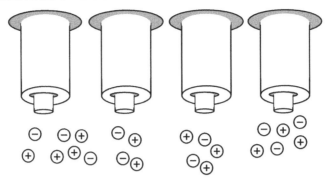

Figure 2.5 Air ionizers

wheels. For good grounding resistive characteristics it is desirable that the resistance from the seat of the chair to the floor ground is within a dissipative resistance range (e.g., a minimum and maximum defined resistance value). A standard for the manufacturing seating is ANSI/ESD STM12.1-2006 Seating- Resistive Measurements [45]. Two tests can be performed to verify compliance of the resistance requirements. A first test can be evaluation of the back of the seating, and seated area, to an equipment ground (with a integrated checker or meter between the chair and the equipment ground point). To contact to the chair region, one electrode is used as specified in the prior sections. To verify conformance, different locations on the chair should be evaluated (e.g., center of the seat panel, center of the seat back, foot ring, and any worn or dirty areas) [23].

2.15 CARTS

Semiconductor parts, trays, card assemblies, and equipment are placed on carts in a manufacturing environment. An operator was careful and placed a wrist strap to prevent charging of the assembly part prior to putting it on the cart. A chain attached to the cart can be noisy as the cart is moved. The operator flipped the chain so it would not disturb others in the manufacturing environment. Another operator took the assembly and pushed it through the facility. As the cart was moved, the operator and the cart began to charge up by charge transfer through the cart wheels. Another operator, who was electrically grounded, removed the assembly from the cart. The cart, and the assembly are charged, and with its removal discharged the assembly.

Figure 2.6 Photo of commercial air ionizers. Permission granted from Transforming Technologies Inc.

2.16 PACKAGING AND SHIPPING

An important part of the ESD control system in manufacturing is proper choice of packaging, and packing materials. The interaction of the packaging materials, and the ESD sensitive parts can interact in tribo-electric charging effects. Hence, after the manufacting process, the packaging and shipping is a key component and culprit of ESD product failures. In 1981, J.M. Kolyer and W.E. Anderson highlighted the importance and proper selection of packaging materials for ESD sensitive items [49]. This work was followed by studies from J.R. Huntsman, B. Unger, D. L. Hart, M.C. Jon, D. Robinson-Hahn, and T.L. Welsher [48,50–52]. Today, there are standards for ESD safe bags, such as ANSI/ESD STM11.31-2006 Bags [53].

2.16.1 Shipping Tubes

ESD tribo-charging can occur within shipping tubes used for shipping of semiconductor components, such as DRAMs. It can be shown that the motion of the chips in a plastic shipping tube can lead to not only charging of the semiconductor chips. Chip motion occurs from the transport, loading and unloading of the components. B. A. Unger, R.G. Chemelli, P.R. Bossard, and M.R. Hudock showed this effect in a publication in 1981 noting that the charging is a function of the shipping tube material [48]. A well-known effect is when the charged chips leave the shipping tube, if they discharge to the work-surface, the corner pins can have ESD failures. It can be demonstrated that the moving charged chips also lead to electromagnetic interference (EMI) noise which is measureable (Figure 2.7).

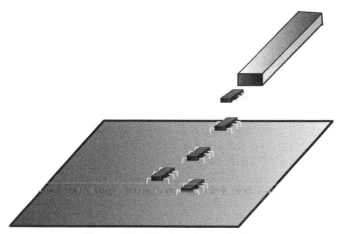

Figure 2.7 Shipping tubes and chips

2.16.2 Trays

Magneto-resistive (MR) heads were placed in shipping trays after MR head dicing. For ESD sensitivity, carbon is extruded into the plastic tray to achieve the correct conductivity. To verify the resistance across the tray, a resistance measurement was made on the two opposite corners of the shipping tray.

MR heads were failing from ESD concerns, leading to a corporate wide investigation. After six months of analysis, it was found that the carbon was not extruded in a uniform fashion in the plastic trays, leading to regional "hot spots" in the trays where the MR heads were failing. Measuring the trays at opposite locations did not provide the correct information to evaluate the carbon uniformity.

To improve the resistance characteristics of shipping trays, carbon is extruded into the plastic to provide the desired resistivity. The individual chips are placed into these trays. The practical problem exists that to verify compliance, how do you guarantee that the trays have the desired resistivity in all sections of the tray? Hence, a method to verify is important. This problem impacted the magnetic recording industry which could not explain failures they were observing. In any future sensitive devices, this may be an issue.

2.17 ESD IDENTIFICATION

A method to provide awareness and identification of ESD sensitive components, assemblies, and systems is usage of symbols and labels. Examples of ESD symbols exist in ANSI/ESD S8.1 – 2007 Symbols – ESD Awareness [54]. A symbol for ESD susceptibility is the ESD Susceptibility Symbol; this symbol can be directly attached to integrated circuits, boards, assemblies, and systems. A second ESD symbol is the ESD Protective Symbol; this indicates ESD protective materials which can be placed on mats, chairs, and other ESD protection supporting equipment that provides ESD protection.

2.18 ESD PROGRAM MANAGEMENT – TWELVE STEPS TO BUILDING AN ESD STRATEGY

In the practical implementation of an ESD program, there are a number of steps to be taken in delivering ESD sensitive parts successfully. In ESD program management, T. Dangelmeyer noted twelve "critical factors" for building an ESD strategy from the product to the customer [24]. These are as follows:

- Effective implementation plan.
- Management commitment.
- A full-time coordinator.

- An active ESD committee.

- Realistic requirements.

- ESD training for measureable goals.

- Auditing using scientific measures.

- ESD test facilities.

- A communication program.

- Systematic planning.

- Human factor engineering.

- Continuous improvement.

The focus of this program is the management of a facility and corporation in managing its staff, tooling and establishing corporate objectives.

2.19 ESD PROGRAM AUDITING

To have a successful ESD program it is important to establish an auditing process to evaluate conformance on establishing the manufacturing environment, and continuous evaluation. There are three types of ESD audits that are practiced today. A first audit type is program management audits, which evaluates implementation plans, requirements, and verification practices. This first audit is typically driven by an ESD team, and management. A second type of audit is geared toward quality process control and statistical process control issues. This second audit type is driven by manufacturing operations personnel on a daily, weekly or monthly basis. The third type of audit is workplace audits that include electrical measurements and satisfying ESD standards. This third type of audit utilizes the ESD measurement equipment and ESD manufacturing standards discussed in Section 2.5 and Section 2.5.1.

Today, there are existing auditing processes, such as ANSI/ESD S20.20-2007 Protection of Electrical and Electronic Parts, Assemblies, and Equipment [22], also known as "S20.20". To have a successful ESD training and auditing process, a program is needed that covers all employees, and comprehensive ESD training can include tutorials, demonstrations, video training, and other educational means.

2.20 ESD ON-CHIP PROTECTION

ESD control is addressed in the manufacturing environment, ESD on-chip protection, and system-level off-chip protection [55–61]. For testing and compliance to ESD objectives, there exists a number of ESD test standard practice and standards to address ESD phenomena on semiconductor components, and systems [62–79].

2.21 SUMMARY AND CLOSING COMMENTS

Electrostatic discharge control in manufacturing environments is intended to provide a control program to prevent yield loss in the manufacturing process. In this chapter, the objective was to provide the reader with a taste of the issues, test methods, standards, and control programs in manufacturing to provide an ESD protected area.

In Chapter 3, the subject will switch to a focus on electrostatic discharge (ESD), electrical overstress (EOS), electromagnetic interference (EMI) and electromagnetic compatibility (EMC).

REFERENCES

1. McAteer (1979) An effective ESD awareness training program. Proceedings of the Electrical Overstress/Electrostatic Discharge (EOS/ESD) Symposium, pp. 1–3.
2. Schnetker, T.R. (1979) Human factors in electrostatic discharge protection. Proceedings of the Electrical Overstress/Electrostatic Discharge (EOS/ESD) Symposium, pp. 122–125.
3. Halperin, S. (1980) Facility evaluation: Isolating environmental ESD problems. Proceedings of the Electrical Overstress/Electrostatic Discharge (EOS/ESD) Symposium, pp. 192–205.
4. McAteer, R.E., Lucas, G.H., and McDonald, A. (1981) A pragmatic approach to ESD problem solving in the manufacturing environment. Proceedings of the Electrical Overstress/Electrostatic Discharge (EOS/ESD) Symposium, pp. 34–39.
5. McFarland, W.Y. (1981) The economic benefits of an effective ESD awareness and control program – an empirical analysis. Proceedings of the Electrical Overstress/Electrostatic Discharge (EOS/ESD) Symposium, pp. 28–33.
6. Frank, D.E. (1981) The perfect "10" – Can you really have one? Proceedings of the Electrical Overstress/Electrostatic Discharge (EOS/ESD) Symposium, pp. 21–27.
7. Euker, R. (1982) ESD in I.C. assembly (a baseline solution). Proceedings of the Electrical Overstress/Electrostatic Discharge (EOS/ESD) Symposium, pp. 142–144.
8. Kirk, W.J. (1982) Uniform ESD protection in a large multi-department assembly plant. Proceedings of the Electrical Overstress/Electrostatic Discharge (EOS/ESD) Symposium, pp. 165–168.
9. Strand, C.J., Tweet, A., and Weight, M.E. (1982) An effective electrostatic discharge protection program. Proceedings of the Electrical Overstress/Electrostatic Discharge (EOS/ESD) Symposium, pp. 145–156.
10. Dangelmayer, G.T. (1983) ESD   How often does it happen? Proceedings of the Electrical Overstress/Electrostatic Discharge (EOS/ESD) Symposium, pp. 1–5.
11. Downing, M.H. (1983) Control implementation and cost avoidance analysis. Proceedings of the Electrical Overstress/Electrostatic Discharge (EOS/ESD) Symposium, pp. 6–11.
12. Hansel, G.E. (1983) The production operator: Weak link or warrior in the ESD battle? Proceedings of the Electrical Overstress/Electrostatic Discharge (EOS/ESD) Symposium, pp. 12–16.
13. Dangelmayer, G.T. (1984) A realistic and systematic ESD control plan. Proceedings of the Electrical Overstress/Electrostatic Discharge (EOS/ESD) Symposium, pp. 1–6.
14. Dangelmayer, G.T. and Jesby, E.S. (1985) Employee training for successful ESD control. Proceedings of the Electrical Overstress/Electrostatic Discharge (EOS/ESD) Symposium, pp. 20–23.
15. Lindholm, A.W. (1985) A case history of an ESD problem. Proceedings of the Electrical Overstress/Electrostatic Discharge (EOS/ESD) Symposium, pp. 10–14.

16. Halperin, S. (1986) Estimating ESD losses in the complex organization. Proceedings of the Electrical Overstress/Electrostatic Discharge (EOS/ESD) Symposium, pp. 12–18.

17. Lai, E. and Plaster, J. (1987) ESD control in the automotive electronic industry. Proceedings of the Electrical Overstress/Electrostatic Discharge (EOS/ESD) Symposium, pp. 10–17.

18. Zezulka, R.J. (1989) Tracking results of an ESD control program. Proceedings of the Electrical Overstress/Electrostatic Discharge (EOS/ESD) Symposium, pp. 36–42.

19. ESD Association ESD-ADV 1.0 2009, *Glossary*, ESD Association, Rome, N.Y.

20. ANSI/ESD S20.20-2007, *Protection of Electrical and Electronic Parts, Assemblies, and Equipment*, ESD Association, Rome, N.Y.

21. ESD TR20.20, 2008, *Handbook*, ESD Association, Rome, N.Y.

22. ANSI/ESD S20.20-2007, *Standard for the Development of Electrostatic Discharge Control Program*, ESD Association, Rome, N.Y.

23. ESD TR53-01-06, 2006, ESD Technical Report for the Protection of Electrostatic Discharge Susceptible Items – Compliance Verification of ESD Protective Equipment and Materials.

24. Dangelmayer, T. (1990) *ESD Program Management: A Realistic Approach to Continuous Measurable Improvement in Static Control*, Kluwer Academic Publishers, New York.

25. ESD ADV 11.2, *Triboelectric Charge Accumulation*, ESD Assocation, Rome, N.Y.

26. ANSI/ESD S6.1, *Grounding*, ESD Association, Rome, N.Y.

27. ANSI/NFPA 70. National Electric Code.

28. ESD DSTM11.13-2009, *Two Point Resistance Measurement*, ESD Association, Rome, N.Y.

29. ANSI/ESD STM11.11-2006, Surface Resistance Measurement of Static Dissipative Planar Materials.

30. ANSI/ESD SP15.1-2005, *In Use Resistance Testing of Gloves and Finger Cots*, ESD Association, Rome, N.Y.

31. ESD STM13.1-2000, *Electrical Soldering/Desoldering Hand Tools*, ESD Association, Rome, N.Y.

32. ESD ADV53.1, *ESD Protective Workstations*, ESD Association, Rome, N.Y. 1995.

33. ANSI/ESD S4.1-2006, *Worksurfaces – Resistance Measurements*, ESD Association, Rome, N.Y.

34. ANSI/ESD STM3.1 – 2006, ESD Protective Worksurfaces - Charge Dissipation Characteristics, Rome, N.Y.

35. ANSI/ESD S1.1-2006, *Wrist Straps*, ESD Association, Rome, N.Y.

36. ESD TR1.0-01-01, *Survey of Constant (Continuous) Monitors for Wrist Straps*, ESD Association, Rome, N.Y. 2001.

37. ANSI/ESD STM9.1-2006, *Footwear – Resistive Characterization*, ESD Association, Rome, N.Y.

38. ESD SP9.2-2003, *Footwear – Foot Grounders Resistive Characterization*, ESD Association, Rome, N.Y.

39. ANSI/ESD STM97.1-2006, *Floor Materials and Footwear – Resistance Measurements in Combination with A Person*, ESD Association, Rome, N.Y.

40. ANSI/ESD S7.1 – 2005, *Resistive Characterization of Materials – Floor Materials*, ESD Association, Rome, N.Y.

41. ANSI/ESD S1.1-2006, *Wrist Straps*, ESD Association, Rome, N.Y.

42. ANSI/ESD STM2.1-1997, *Garments*, ESD Association, Rome, N.Y.

43. ANSI/ESD STM3.1 – 2006, *Ionization*, ESD Association, Rome, N.Y.

44. ANSI/ESD SP3.3-2006, *Periodic Verification of Air Ionizers*, ESD Association, Rome, N.Y.

45. ANSI/ESD STM12.1-2006, *Seating- Resistive Measurements*, ESD Association, Rome, N.Y.

46. ANSI/ESD STM4.1-2006, *Worksurfaces – Resistance Measurements*, ESD Association, Rome, N.Y.

47. ANSI/ESD SP10.1-2007, *Automatic Handling Equipment (AHE)*, ESD Association, Rome, N.Y.

48. Unger, B.A., Chemelli, R.G., Bossard, P.R., and Hudock, M.R. (1981) Evaluation of integrated circuit shipping tubes. Proceedings of the Electrical Overstress/Electrostatic Discharge (EOS/ESD) Symposium, pp. 57–64.

49. Kolyer, J.M. and Anderson, W.E. (1981) Selection of packaging materials for electrostatic discharge sensitive (ESDS) items. Proceedings of the Electrical Overstress/Electrostatic Discharge (EOS/ESD) Symposium, pp. 75–84.

50. Huntsman, J.R. (1984) Triboelectric charge: Its ESD ability and a measurement method for its propensity on packaging materials. Proceedings of the Electrical Overstress/Electrostatic Discharge (EOS/ESD) Symposium, pp. 64–77.

51. Unger, B.A. and Hart, D.L. (1985) Triboelectric characterization of packaging materials. Proceedings of the Electrical Overstress/Electrostatic Discharge (EOS/ESD) Symposium, pp. 107–110.

52. Jon, M.C., Robinson-Hahn, D., and Welsher, T.L. (1988) Tape and reel packaging – an ESD concern. Proceedings of the Electrical Overstress/Electrostatic Discharge (EOS/ESD) Symposium, pp. 15–23.

53. ANSI/ESD STM11.31-2006, *Bags*, ESD Association, Rome, N.Y.

54. ANSI/ESD S8.1 – 2007 Symbols – ESD Awareness.

55. Voldman, S. (2004) *ESD: Physics and Devices*, John Wiley and Sons, Ltd., Chichester, England.

56. Voldman, S. (2005) *ESD: Circuits and Devices*, John Wiley and Sons, Ltd., Chichester, England.

57. Voldman, S. (2006) *ESD: RF Circuits and Technology*, John Wiley and Sons, Ltd., Chichester, England.

58. Voldman, S. (2008) *ESD: Circuits and Devices*, Publishing House of Electronic Industry (PHEI), Beijing, China.

59. Voldman, S. (2009) *ESD: Failure Mechanisms and Models*, John Wiley and Sons, Ltd., Chichester, England.

60. Voldman, S. (2007) *Latchup*, John Wiley and Sons, Ltd., Chichester, England.

61. Ker, M.D. and Hsu, S.F. (2009) *Transient Induced Latchup in CMOS Integrated Circuits*, John Wiley and Sons, Ltd., Singapore.

62. ANSI/ESD ESD-STM 5.1 – 2007 (2007) ESD Association Standard Test Method for the Protection of Electrostatic Discharge Sensitive Items - Electrostatic Discharge Sensitivity Testing - Human Body Model (HBM) Testing -Component Level. Standard Test Method (STM) document.

63. ANSI/ESD ESD-STM 5.2 – 1999 (1999) ESD Association Standard Test Method for the Protection of Electrostatic Discharge Sensitive Items - Electrostatic Discharge Sensitivity Testing - Machine Model (MM) Testing -Component Level. Standard Test Method (STM) document.

64. ANSI/ESD ESD-STM 5.3.1 – 1999 (1999) ESD Association Standard Test Method for the Protection of Electrostatic Discharge Sensitive Items - Electrostatic Discharge Sensitivity Testing – Charged Device Model (CDM) Testing -Component Level. Standard Test Method (STM) document.

65. ESD Association DSP 14.1-2003 (2003) ESD Association Standard Practice for the Protection of Electrostatic Discharge Sensitive Items – System Level Electrostatic Discharge Simulator Verification Standard Practice. Standard Practice (SP) document.

66. ESD Association DSP 14.3-2006 (2006) ESD Association Standard Practice for the Protection of Electrostatic Discharge Sensitive Items – System Level Cable Discharge Measurements Standard Practice. Standard Practice (SP) document.

67. ESD Association DSP 14.4-2007 (2007) ESD Association Standard Practice for the Protection of Electrostatic Discharge Sensitive Items – System Level Cable Discharge Test Standard Practice. Standard Practice (SP) document.

68. Geski, H. (September 2004) DVI compliant ESD protection to IEC 61000-4-2 level 4 standard. *Conformity*, 12–17.

69. ANSI/ESD Association ESD-SP 5.5.1-2004 (2004) ESD Association Standard Practice for the Protection of Electrostatic Discharge Sensitive Items - Electrostatic Discharge Sensitivity Testing – Transmission Line Pulse (TLP) Testing Component Level. Standard Practice (SP) document.

70. ANSI/ESD Association ESD-STM 5.5.1-2008 (2008) ESD Association Standard Test Method for the Protection of Electrostatic Discharge Sensitive Items - Electrostatic Discharge Sensitivity Testing – Transmission Line Pulse (TLP) Testing Component Level. Standard Test Method (STM) document.

71. ESD Association ESD-SP 5.5.2 (2007) ESD Association Standard Practice for the Protection of Electrostatic Discharge Sensitive Items - Electrostatic Discharge Sensitivity Testing Very Fast Transmission Line Pulse (VF-TLP) Testing Component Level. Standard Practice (SP) document.

72. ANSI/ESD Association ESD-SP 5.5.2-2007 (2007) ESD Association Standard Practice for the Protection of Electrostatic Discharge Sensitive Items - Electrostatic Discharge Sensitivity Testing – Very Fast Transmission Line Pulse (VF-TLP) Testing Component Level. Standard Practice (SP) document.

73. ESD Association ESD-STM 5.5.2 (2009) ESD Association Standard Test Method for the Protection of Electrostatic Discharge Sensitive Items - Electrostatic Discharge Sensitivity Testing Very Fast Transmission Line Pulse (VF-TLP) Testing Component Level. Standard Test Method (STM) document.

74. International Electro-technical Commission (IEC) IEC 61000-4-2 (2001) Electromagnetic Compatibility (EMC): Testing and Measurement Techniques – Electrostatic Discharge Immunity Test.

75. IEC 61000-4-2 (2008) Electromagnetic Compatibility (EMC) – Part 4-2:Testing and Measurement Techniques – Electrostatic Discharge Immunity Test.

76. Chundru, R., Pommerenke, D., Wang, K. *et al.* (2004) Characterization of human metal ESD reference discharge event and correlation of generator parameters to failure levels – Part I: Reference Event. *IEEE Transactions on Electromagnetic Compatibility*, **46** (4), 498–504.

77. Wang, K., Pommerenke, D., Chundru, R. *et al.* (2004) Characterization of human metal ESD reference discharge event and correlation of generator parameters to failure levels – Part II: Correlation of generator parameters to failure levels. *IEEE Transactions on Electromagnetic Compatibility*, **46** (4), 505–511.

78. ESD Association ESD-SP 5.6-2008 (2008) ESD Association Standard Practice for the Protection of Electrostatic Discharge Sensitive Items - Electrostatic Discharge Sensitivity Testing – Human Metal Model (HMM) Testing Component Level. Standard Practice (SP) document.

79. ANSI/ESD SP5.6-2009 (2009) Electrostatic Discharge Sensitivity Testing - Human Metal Model (HMM) - Component Level.

3 ESD, EOS, EMI, EMC and Latchup

A new computer was to be announced with a brand new high density memory chip inside. Wine and cheese was being served to the invited guests prior to the announcement of the new personal computer. When the system was first powered up, a good ground connection was not established. The signal pins on the memory chip was "low" but the substrate ground connection was improperly sequenced. Current flowed through the ESD networks in the chip, leading to "latchup" of the off-chip driver circuitry. Smoke came out of the machine as the semiconductor chip went into an over-current event.

The customers in the other room started to sniff, and ask "where is that smoke coming from?"

3.1 ESD, EOS, EMI, EMC AND LATCHUP

As an introduction to these issues, the chapter will first provide a short description of these subjects. This will be followed by an introduction to the various ESD sources and models, followed by an introduction to EOS issues, and the other areas (Figure 3.1).

3.1.1 ESD

Electrostatic discharge (ESD) is a subclass of electrical overstress and may cause immediate device failure, permanent parameter shifts and latent damage causing increased degradation rate. It has at least one of three components: localized heat generation, high current density and high electric field gradient, and prolonged presence of currents of several amperes transfering energy to the device structure to cause damage.

ESD Basics: From Semiconductor Manufacturing to Product Use, First Edition. Steven H. Voldman.
© 2012 John Wiley & Sons, Ltd. Published 2012 by John Wiley & Sons, Ltd.

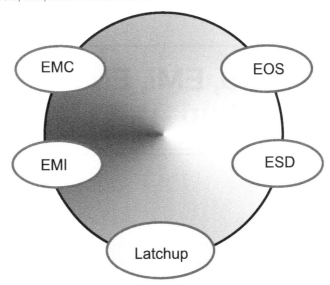

Figure 3.1 ESD, EOS, EMI, EMC, and Latchup

Electrostatic discharge (ESD) is addressed on semiconductor components through ESD circuits, chip architecture and design [1–13]. During ESD events, ESD failure mechanisms occur in the semiconductor devices [1–24]. In ESD semiconductor chip design, the ESD design discipline is customized to different application spaces, such as ESD digital design [1–5,7,9,10], ESD radio frequency (RF) design [6,25], and ESD analog design [13]. With semiconductor component scaling, and both evolutionary and revolutionary changes, ESD devices and design must also evolve [26–31].

ESD test practices have evolved for both components and systems over the last thirty years. ESD standard practices and standards have evolved with the changes of semi-conductor components and new issues [32–51]. For systems, new issues such as cable discharge events [52–59], IEC 61000-4-2 system events [60–63] and human metal model [64–67] have occurred.

3.1.2 EOS

Electrical overstress (EOS) is a wide classification for over-current conditions for electronic components and electronic systems. EOS events can lead to loss of functionality, thermal failure and destruction of electronic components and systems. EOS and ESD are important issues for power and analog semiconductor components [13,68–80].

3.1.3 EMI

Electromagnetic interference (EMI) is interference, or noise, generated from an electromag-netic field. EMI can lead to both component level or system level failure of electronic systems. EMI can lead to failure of electronic components, without physical contact with the electronic system [81–103].

3.1.4 EMC

Electromagnetic compatibility (EMC) is the ability of an electronic system to function properly in its intended electromagnetic environment and not be a source of electronic emissions to that electromagnetic environment [81–103]. Electromagnetic compatibility (EMC) has two features. A first feature is a source of emission of an electromagnetic field. A second feature is the collector of electromagnetic energy. The first aspect is the emission of an electromagnetic field which may lead to electromagnetic interference of other components or systems. The second aspect has to do with susceptibility of a component, or system to the undesired electromagnetic field.

3.1.5 Latchup

Latchup is a term used to describe a particular type of short circuit which can occur in semiconductor components [8,11]. A parasitic structure is formed which consists of a p-channel MOSFET and n-channel MOSFET transistor, leading to a parasitic PNPN structure. An inadvertent low-impedance path between the power supply rails of a MOSFET circuit occurs, leading to a low voltage high current state. This leads to disruption of functionality, and can lead to thermal runaway, electrical overstress and package destruction.

3.2 ESD MODELS

In the evolution of electrostatic discharge (ESD) development over the last thirty years, new ESD simulation models are being introduced. Figure 3.2 shows a number of ESD models being practiced today.

Figure 3.3 shows the evolution of the ESD models, and new models being proposed in the future. The ESD models include the human body model (HBM) [32–38], the machine model (MM) [39], charged device model (CDM) [40], transmission line pulse (TLP) method [41–44], the very-fast transmission line pulse (VF-TLP) method [45–50], as well as the cassette model (also known as the small charge model (SCM).

3.2.1 Human Body Model (HBM)

A fundamental model used in the ESD industry is known as the human body model (HBM) pulse [32–38]. The model was intended to represent the interaction between the electrical discharge from a human being, who is charged, with a component, or object. The model assumes that the human being is the initial condition. The charged source then touches a component or object using a finger. The physical contact between the charged human being and the component or object allows for current transfer between the human being and the

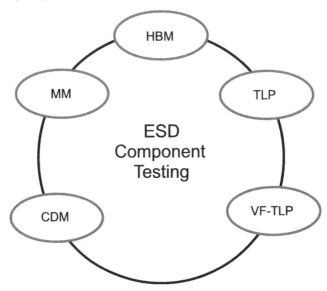

Figure 3.2 ESD models

object. A characteristic time of the HBM event is associated with the electrical components used to emulate the human being. In the HBM standard, the circuit component to simulate the charged human being is a 100 pF capacitor in series with a 1500 Ω resistor. This network has a characteristic rise time and decay time. The characteristic decay time is associated with the time of the network

$$\tau_{HBM} = R_{HBM}C_{HBM}$$

where R_{HBM} is the series resistor and C_{HBM} is the charged capacitor. This is a characteristic time of the charged source. Figure 3.4 shows the human body model (HBM) pulse waveform.

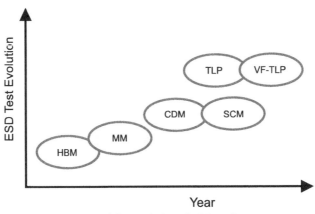

Figure 3.3 Evolution of ESD testing

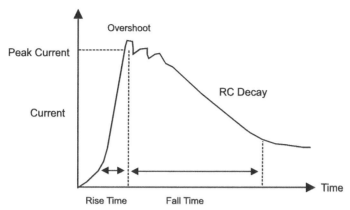

Figure 3.4 Human body model (HBM) pulse waveform

Figure 3.5 shows the equivalent circuit model. The equivalent circuit model includes a capacitor and resistor element. In the HBM standard, the circuit component to simulate the charged human being is a 100 pF capacitor in series with a 1500 Ω resistor.

Figure 3.6 shows a human body model (HBM) source used in an automated HBM test equipment. The source contains a 100 pF capacitor in series with a 1500 Ω resistor.

3.2.2 Machine Model (MM)

Another fundamental model used in the semiconductor industry is known as the machine model (MM) pulse [39]. The MM event was intended to represent the interaction between the electrical discharge from a conductive source, which is charged, with a component, or object. The model assumes that the "machine" is charged as the initial condition. The charged source then touches a component or object. In this model, an arc discharge is assumed to occur between the source and the component or object allowing for current transfer between the charged object and the component or object. A MM characteristic time is associated with the electrical components used to emulate the discharge process. In the

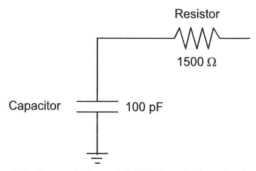

Figure 3.5 Human body model (HBM) equivalent circuit model

Figure 3.6 Photograph of HBM pulse source

MM standard, the circuit component is a 200 pF capacitor with no resistive component (Figure 3.7). An arc discharge fundamentally has a resistance in the order of 10 to 25 Ω. The characteristic decay time is associated with the time of the network

$$\tau_{MM} = R_{MM}C_{MM}$$

where R is the arc discharge resistor and C is the charged capacitor. This is a characteristic time of the charged source.

Figure 3.8 shows an example of the machine model (MM) pulse waveform. Without a large resistor element, the MM pulse waveform is a weak damped oscillation, whose waveform oscillates from a positive to negative polarity. Additionally, the peak current of the MM pulse waveform is significantly higher than a HBM pulse waveform. It is the feature of higher peak current, as well as polarity transitions which makes this ESD test more difficult to achieve the desired specification objectives. Figure 3.9 is the MM source from an automated MM ESD tester.

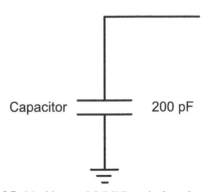

Figure 3.7 Machine model (MM) equivalent circuit model

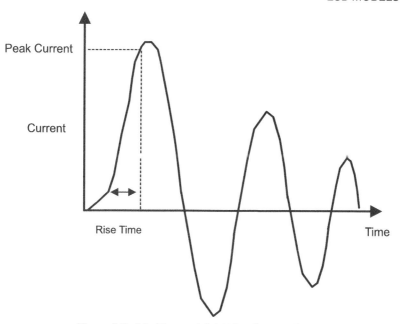

Figure 3.8 Machine model (MM) pulse waveform

Figure 3.9 Machine model (MM) source

3.2.3 Cassette Model

The Cassette Model (CM), also known as the Charged Cassette Model (CCM), and small charge model (SCM) are a recent model associated with consumer electronics. In consumer electronics there are many applications where a human plugs a small cartridge or cassette into an electronic socket. These are evident in popular electronic games. The cassette model is of interest to corporations in the "game industry." In today's electronic world, there are many palm size electronic components which must be socketed into a system for non-wireless applications. To verify the electronic safety of such equipment, the cassette itself is assumed as a charged source. The "cassette model" assumes a small capacitance, and negligible resistance. This model is equivalent to a MM-type current source with a much lower capacitor component. The model assumes the resistance of an arc discharge, and a capacitance of 10 pF [10]. In the CCM, customers' objectives for this model are in the order of 600 V.

3.2.4 Charged Device Model (CDM)

A quality and reliability team wanted to start requiring the "charged device model" standard as part of the qualification. The design team, who did not want to risk the release of their ASIC library refused to do it. The Quality and Reliability team gave up the fight.

The qualification test vehicle with all the circuitry for the qualification was manufactured, diced and then packaged overseas in Japan. When the parts were returned, there were four pins that were failing – two on the top, and two on the bottom. One hundred percent of the hardware had this failure, and the circuit design qualification failed due to lack of hardware. The two top pins and two bottom pins were the area of a manufacturing person's finger.

Figure 3.10 is a chart of the CDM testing process. In the CDM process, a semiconductor chip is placed "bug up" with the backside of the semiconductor chip on an insulating surface [40]. The semiconductor chip is charged through the ground pin. The ground chip is removed, leaving the chip fully charged on the insulating surface. A pogo pin is then placed over a signal pad, and dropped onto the pin (or bond pad, or solder ball) where the semiconductor chip is discharged through to a 1 Ω ground connection.

Figure 3.11 is an example of a commercial CDM tester used in the qualification of semiconductor chips. The commercial CDM testers are designed in compliance with the CDM test standards [40].

3.2.5 Transmission Line Pulse (TLP)

Transmission line pulse (TLP) testing has seen considerable growth in the ESD discipline [41–44]. In this form of ESD testing, a transmission line cable is charged using a voltage source. The TLP system discharges the pulse into the device under test (DUT). The characteristic time of the pulse is associated with the length of the cable. The pulse width of a transmission line pulse is a function of the length of the transmission line and the propagation velocity of the transmission line.

TLP systems are designed in different configurations. TLP system configurations include current source, Time Domain Reflectometry (TDR), Time Domain Transmission (TDT), and Time Domain Reflectometry and Transmission (TDRT) [41]. In all configurations, the source

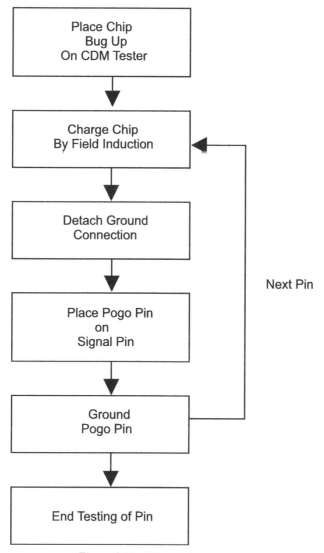

Figure 3.10   CDM testing process

is a transmission line whose characteristic time constant is determined by the length of the transmission line cable. The various TLP configurations influence the system characteristic impedance, the DUT location, and the measurement of the transmitted or reflected signals. For this method, the choice of pulse width is determined by the interest in using TLP testing as an equivalent or substitute method to the HBM methodology. It is standard practice today, to choose the TLP cable length to provide a TLP pulse width of 100 ns with less than 10 ns rise time. Figure 3.12 is an example of the TLP waveform.

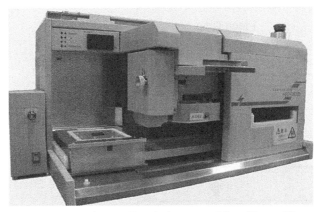

Figure 3.11 CDM test equipment. Permission granted from Hanwa Electronics, Inc.

Figure 3.13 is an example of the TLP system in a Time Domain Reflectometry (TDR) test configuration [41–44]. In this configuration, there is an incident wave, and a reflected wave. To determine the absorbed power in the device, the absorbed power equals the incident power minus the reflected power. From the incident and reflected wave, the current and voltage across the device under test (DUT) can be evaluated.

Figure 3.14 is an example of a transmission line pulse (TLP) test system in a Time Domain Transmission (TDT) test configuration [41–44]. In this configuration, there is an incident wave, and a transmitted wave. To determine the absorbed power in the device, the absorbed power equals the incident power minus the transmitted power. ESD design engineers are mostly interested in the TLP I-V pulsed waveform which is constructed from obtaining the current and voltage across the device under test (DUT).

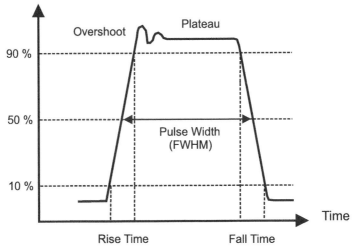

Figure 3.12 Transmission line pulse model – pulse waveform

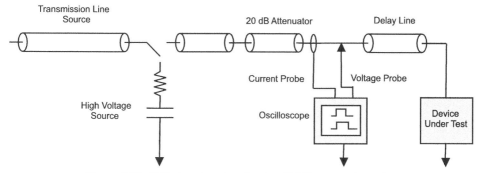

Figure 3.13 TLP time domain reflection (TDR) test configuration

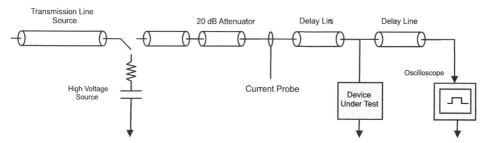

Figure 3.14 TLP time domain transmission (TDT) test configuration

Figure 3.15 is a transmission line pulse (TLP) test system in a Time Domain Reflectometry and Transmission (TDRT) test configuration [41–44]. In this configuration, there is an incident wave, a reflected wave, and a transmitted wave. To determine the absorbed power in the device, the absorbed power equals the incident power minus the transmitted power, and the reflected power.

Figure 3.16 is an example of a TLP pulsed current-voltage (I-V) plot. In this plot, each data point represents a separate pulse applied to the device. The pulse width is fixed, as the magnitude of the current is increased. The voltage and current are measured across the device under test (DUT), and the pulsed I-V plot is constructed.

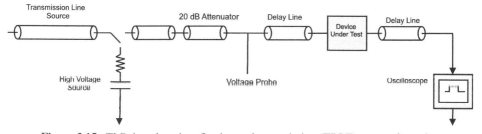

Figure 3.15 TLP time domain reflection and transmission (TDRT) test configuration

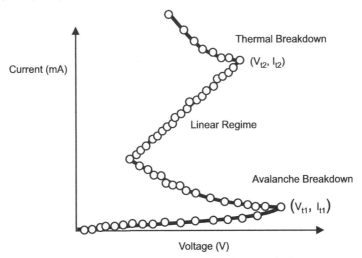

Figure 3.16   TLP pulsed current-voltage (I-V) plot

3.2.6 Very Fast Transmission Line Pulse (VF-TLP)

The very fast TLP (VF-TLP) test method is similar to the TLP methodology [45–50]. The interest in VF-TLP is driven by understanding semiconductor devices in a time regime similar to the CDM time constant. The characteristic time of interest is again determined by the propagation characteristics of the transmission line cable source, and the length of the transmission line cable. VF-TLP pulse width of interest is a pulse width of less than 5 nanoseconds and a sub-1 ns rise time. This time regime is well below the thermal diffusion time constant in semiconductor media. The method of the fast time constant limits the acceptable configurations of the VF-TLP system and suitable equipment for measurement.

3.3 ELECTRICAL OVERSTRESS (EOS)

Electrical overstress (EOS) is a concern in semiconductor components to electronic systems. EOS events are one of the classifications of electrical conditions observed.

EOS events can include the following:

- Electrostatic Discharge (ESD).
- Lightning.
- System transients.
- Electromagnetic pulse (EMP).
- Charging.

ESD is a subset of EOS phenomena. Electrical overstress can also include lightning, system transients, electromagnetic pulse (EMP), and forms of charging. Charging can occur in a wide range of events in semiconductor chip manufacturing, product assembly, system assembly to application usage. Wafer-level semiconductor charging can occur due to the following:

- Plasma arcing.

- Semiconductor etch.

- Wafer dicing.

- Wafer carrying devices (e.g., boxes, trays, and shipping vehicles).

Semiconductor chips can undergo charging effects in shipping tubes, and trays prior to packaging.

In this chapter, the focus will be on EOS events associated with over-current and over-voltage events that occur in semiconductor chips, packages, and systems.

3.3.1 EOS Sources – Lightning

Lightning is a source of electrical overstress (EOS) which is a concern for telecommunications and aircraft. Lightning strikes or indirect lightning effects have many standards used in the industry. Examples of standards to address lightning are as follows:

- RTCA/DO-160E Section 22.

- SAE ARP5412.

- EUROCAE/ED-14E.

- FAA AC:20-136.

- SAE AE4L.

- MIL-STD-1757.

EOS test combinations include the following procedures:

- Pin injection.

- Cable injection.

- Ground injection.

These EOS standards address different forms of lightning events:

- Single stroke lightning.

- Multiple stroke lightning.

- Multiple bursts.

For the EOS test combinations, the pin injection, cable injection or ground injection can use single stroke, multiple stroke or multiple burst events. In each of these conditions, there exists Waveform Sets and Test Levels. The Waveform Sets are formed from a combination of specified waveforms. These Waveform Sets and Test Levels are defined in some of the above standards, such as RTCA DO 160 E Section 22. The European Standard EUROCAE/ED-14D is harmonized with the RTCA DO 160 test standard.

3.3.2 EOS Sources – Electromagnetic Pulse (EMP)

A source of an electrical overstress (EOS) event is the Electromagnetic Pulse (EMP) event. The electromagnetic pulse (EMP) event is defined as a high amplitude, single pulse, short duration, and broadband pulse of electromagnetic energy. EMP is the electromagnetic effect resulting from the detonation of a nuclear device. Other definitions of EMP at high altitudes are HEMP, HAEMP and HNEMP.

EMP test methods include MIL-STD-461; this provides radiated (RS 105) and conducted (CS 116) test methods and test levels for determining a device's immunity to EMP. The RS 105 radiated test methodology addresses the risk of radiated exposure to an EMP event; this test is applicable to equipment installed in exposed and partially exposed environments on aircraft, surface ships, submarines, and ground vehicles. The CS116 coupling test method addresses the effects of EMP coupling onto interconnecting wiring harnesses; this test is to ensure the equipment's immunity to damped sinusoidal transients induced on the equipment's cables. Testing is generally the norm in all applications with limited applicability to submarine equipment.

The MIL-STD-188-125 establishes minimum requirements and design objectives for high-altitude electromagnetic pulse (HAEMP) hardening of fixed, ground-based facilities that perform critical, time-urgent command, control, communications, computer, and intelligence (C4I) missions. Similar to the approach described in MIL-STD-461, this standard provides both radiated and conducted test methods and test levels.

3.3.3 EOS Sources – Machinery

Machinery can be a source of electrical overstress when it is contained on the same power lines as other equipment. Pumps, motors, and other elements are large inductive loads that lead to significant transients, and inductive interaction of the load, and other components attached. Electrical components on the same circuit and same ground will be influenced by switching of machinery.

3.3.4 EOS Sources – Power Distribution

Power distribution can be a source of electrical overstress (EOS). Power distributions have both high currents, power surges, and other forms of load transitions. In an electrical system, sequencing of the power up, and power down can lead to electrical current overload. An example of sequencing issues can lead to forward biasing of ESD protection networks or off-chip driver (OCD) circuitry. Even on a semiconductor chip, the switching of power transistors can influence the low voltage circuitry and analog circuits on the same substrate wafer. LDMOS devices are vulnerable to EOS failures due to the high voltages and power transients on the same circuit domain.

3.3.5 EOS Sources – Switches, Relays and Coils

Switches, relays and coils can act as sources of electrical overstress. These elements can be the source for high transient currents, noise, and inductive load transitions. Switches introduce high current transients during "turn-off" or "turn-on." Switches can be CMOS circuitry, or power LDMOS transistors in a "switch" configuration. Relays and coils introduce inductive loads, that can lead to voltage transients when there is a rapid change in current through these elements.

3.3.6 EOS Design Flow and Product Definition

In the development of an EOS robust component, the design flow deviates from a standard product design flow. In the product design flow, the process includes the evaluation of the market, establishing the specifications, defining cost estimates, developing a design team, and a schedule. The standard products undergo standard test requirements and reliability requirements. Figure 3.17 shows the standard process flow for a product.

For EOS robust components, a greater synthesis of the environment and technology definition is required. For EOS robust components, the design flow considerations are as follows:

- Market definition.

- Environment.

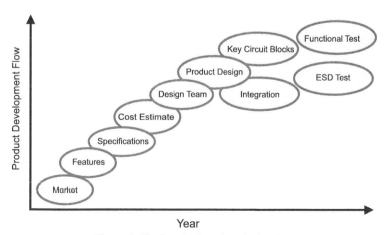

Figure 3.17 Standard product design flow

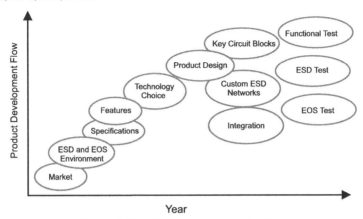

Figure 3.18 EOS environment design flow

- Application Features and Specifications.

- Technology Conformance to Application and Specification.

- Technology Choice and Definition.

- EOS and ESD Device Design and Definition.

For the EOS robust components, the market definition needs to be understood. One of the most important issues is an understanding of the environment in which the component will be placed. Figure 3.18 is a product design flow where the EOS environment is addressed at the beginning of the product definition stage.

3.3.7 EOS Sources – Design Issues

Many of the electrical overstress (EOS) issues can occur from the design of the semiconductor component, the system and its integration. Examples of EOS source design issues are as follows (Figure 3.19):

- Semiconductor process – application mismatch.

- Printed circuit board (PCB) inductance.

- Printed circuit board (PCB) resistance.

- Latchup sensitivity.

- Safe operating area (SOA) power rating violation.

- Safe operating area (SOA) voltage rating violation.

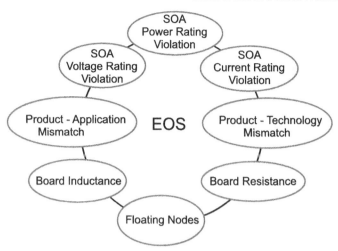

Figure 3.19 EOS design issues

- Safe operating area (SOA) current rating violation.
- Transient safe operating area – di/dt and dv/dt.

3.3.8 EOS Failure Mechanisms

EOS failure mechanisms can occur from the aforementioned issues. They can result from technology incompatibility, boards, loads and integration of systems.

3.3.8.1 *EOS Failure Mechanisms – Semiconductor Process – Application Mismatch*

Electrical overstress can be a concern when the semiconductor process is incompatible with the application. One of the key problems in the semiconductor industry is that the developers of the semiconductor technology have specific design applications in mind for potential customers. It is very common for chip development teams to utilize technologies for applications that were not the intention of the semiconductor foundry. As a result, the semiconductor technology high voltage characteristics are not suitable for many applications.

3.3.8.2 *EOS Failure Mechanisms – Bond Wire Failures*

Bond wire failure can occur from EOS events due to the current magnitude exceeding the bond wire reliability. Bond wires can carry current magnitudes in the order of 1 A to 1.5 A. When these current magnitudes are exceeded, failure of the bond wire can occur.

Sources of bond wire failure can be as follows:

• Reverse insertion of component.

• Over-heating or self-heating.

• Latchup.

• Bond wires inadequate of transient power requirements.

• Bond wire mis-match.

Reverse insertion of the component can lead to failure of the bond wire. Reverse insertion can lead to forward biasing of semiconductor elements or over-voltage of internal component elements.

Internal heating and over-heating can lead to bond wire failure. Power devices can have significant dc as well as transient Joule heating which can impact bond wire lifetime.

CMOS latchup can also lead to excessive current magnitudes above the specification of the bond wire. CMOS latchup can lead to a thermal runaway event, and component failure. Current magnitudes can lead to chip failure, bond wire, and packaging failure.

In power applications, multiple bond wires are placed for a given power supply or ground. In the case of multiple bond wires, if there is an impedance mismatch, the current in each of the bond wires may not be identical. This can lead to a bond wire mis-match where one bond wire current magnitude is exceeded.

Bond wire failure can be observed using x-ray analysis of a component.

3.3.8.3   EOS Failure Mechanisms – External Load to Chip Failures

Electrical overstress failure can occur from the presence of external loads whose interaction with the semiconductor chip was not anticipated. External loads that are a common issue are as follows:

• Capacitance loads.

• Inductive loads.

Capacitance loads can be a concern during power up, or power down of a component.

An example of this is a linear regulator. For a regulator application, if the output capacitance is too large during power up, a surge current can occur on the output of charging capacitor. The large surge current from the semiconductor can lead to failure of the semiconductor chip power rails, interconnects, and bond wires.

For inductive loads, unanticipated interactions can occur between the chip and the external inductive load. The inductive load can be initiated by the printed circuit board or an external inductor load. As an example, the RC-triggered ESD power clamp can interact with the inductive load, leading to an RLC response, instead of the intended RC response of the on-chip protection circuitry.

3.3.8.4   EOS Failure Mechanisms – Printed Circuit Board (PCB) to Chip Failures

A semiconductor chip was tested for electrical overstress as a separate component, and passed at the semiconductor plant qualification. When the component was mounted on the printed circuit board, the capacitance of the board lead to a "Miller capacitance" coupling of the off-chip driver (OCD), lead to card failure and an EOS event. A 4% yield fallout for card failure was apparent in the customers' system.

Electrical overstress (EOS) can occur due to printed circuit boards, and interaction of the chips mounted on the printed circuit board. Examples of this issue are as follows:

• Printed Circuit Board (PCB): Large inductance of printed circuit board (PCB).

• PCB-Chip Interaction: Capacitance increase on signal pins due to card.

3.3.8.5   EOS Failure Mechanisms – Reverse Insertion

Reverse insertion can lead to EOS failure of components. Semiconductor chips that are placed on a board where the V_{DD} and V_{SS} are reversed can lead to failure. In CMOS technology, the p-channel MOSFET source is normally connected to the V_{DD} power supply, and the n-channel MOSFET source is normally connected to the V_{SS} power supply. If the chip is inserted opposite to its intended application, the CMOS inverter circuits can lead to a high current, and can lead to bond wire failures.

3.4 EMI

Electromagnetic interference (EMI) is electromagnetic phenomena that effects both components and systems [81–103]. Electromagnetic interference has not been a major problem internal to semiconductor components to date, but it is anticipated that with the device scaling from microelectronics to nanoelectronics this concern will be of greater importance. Today, EMI is a concern in unprotected components, such as those in magnetic recording devices. Electromagnetic fields can establish a differential voltage across a magneto-resistor (MR) component leading to functional failures. As devices are scaled from MR heads to giant magneto-resistors (GMR), and tunneling magneto-resistors (TMR), the sensitivity to EMI increases.

In system on chip (SOC) applications, signal lines between digital and analog domains can observe EMI sensitivity. Additionally, radio frequency (RF) circuits contained on mixed signal chips (e.g., digital, analog, and RF) can also sense EMI noise from the digital circuitry. One technique to eliminate this effect is to place an EMI shield around the RF circuits.

3.5 EMC

Electromagnetic compatibility is a concern in electronic systems [81–103]. Electronic systems are susceptible to noise due to external sources, or sources within a card, board or semiconductor chip. EMC will be discussed in Chapter 4.

3.6 LATCHUP

> *A corporation wanted to start doing CMOS development for the first time. Although
> the RCA "COSMOS" technology (later called CMOS) was practiced for space appli-
> cations, few corporations attempted this new technology for terrestrial earth bound
> applications. Researchers from the corporation developed a technology that con-
> tained both the p-channel MOSFET device and n-channel MOSFET device for the first
> time. The ground rules for this new technology were aggressive to make a competitive
> and dense technology. The technology was called "Beacon." It was never produced
> since it underwent CMOS latchup, and the corporation management said that they
> would not practice this CMOS technology again!*

Latchup occurs in semiconductor technology as a result of the presence of a parasitic pnp
and npn bipolar transistor forming a pnpn (e.g., silicon controlled rectifier) [8,11]. Latchup
robustness is a strong function of semiconductor technology in CMOS and BiCMOS technology. The electrical, thermal, and spatial connectivity can influence the latchup robustness of a semiconductor chip. Latchup can be both a local as well as global phenomena in a
semiconductor chip. The electrical connectivity of pads, circuits, and circuit functional
blocks can influence the latchup robustness of a semiconductor chip. The placement of
pads, circuits, circuit function, power rails, and the architecture of the semiconductor chip
or system can also play a key role in the latchup sensitivity. As with noise, the relative
placement of injection sources, and latchup-sensitive circuits is key to latchup robustness.
Analogous to the noise issue, semiconductor process choices, and semiconductor layout
design influences latchup sensitivity. Substrate contacts, isolation structures, and the relative placement of circuits influence the spatial and electrical coupling which occurs in the
semiconductor chip. Hence the spatial and electrical solutions can influence noise, latchup
and ESD protection. Latchup design practices, as is true in ESD design practice, can
involve the following concepts:

- Addressing high current and high voltage sources.
- Involvement of parasitic elements.
- Utilization of guard rings.
- De-coupling from power supply and ground connections.

Latchup is a fundamental issue in semiconductor chip design which is addressed with
both process, and design solutions, as well as ground rules, checking systems and verification
[8,11]. Figure 3.20 highlights some of the latchup interactions of concern.

> *In early CMOS development, a semiconductor chip package exploded as a part was
> placed in the application. Management never saw anything like this before; so they
> set up a corporate task force called the "Exploding Plastic Task Force." All kinds of
> corporate experts were invited, from package engineers, plastic engineers, bond,
> assembly, and one semiconductor device engineer. His expertise was CMOS Latchup.*

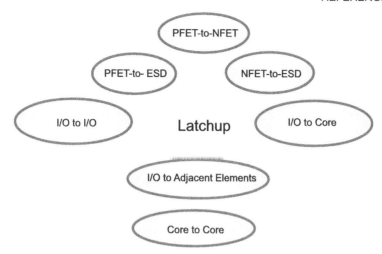

Figure 3.20 Latchup design issues

A space mission had high speed communication between its planetary rover and the earth. A single heavy ion particle triggered "Single event Latchup (SEL)" in the mission. Fortunately, a secondary redundant communication system was in the system to allow continued success of the space mission.

3.7 SUMMARY AND CLOSING COMMENTS

In this chapter, we provided an introduction to electrostatic discharge (ESD), electrical overstress (EOS), electromagnetic interference (EMI), electromagnetic compatibility (EMC), and latchup. Each of these fields has a vast number of publications, literature, and books. In our introduction, we have provided some of the language, terms, and testing standards.

In Chapter 4, discussion is continued on ESD to EMC issues in systems. After this chapter, we will return to a discussion on solving the problems in both components to systems.

REFERENCES

1. Dabral, S. and Maloney, T.J. (1998) *Basic ESD and I/O Design*, John Wiley and Sons Ltd., West Sussex.
2. Wang, A.Z.H. (2002) *On Chip ESD Protection for Integrated Circuits*, Kluwer Publications, New York.
3. Amerasekera, A. and Duvvury, C. (2002) *ESD in Silicon Integrated Circuits*, 2nd edn, John Wiley and Sons, Ltd., West Sussex.
4. Voldman, S. (2004) *ESD: Physics and Devices*, John Wiley and Sons, Ltd., Chichester, England.
5. Voldman, S. (2005) *ESD: Circuits and Devices*, John Wiley and Sons, Ltd., Chichester, England.
6. Voldman, S. (2006) *ESD: RF Circuits and Technology*, John Wiley and Sons, Ltd., Chichester, England.

7. Voldman, S. (2009) *ESD: Failure Mechanisms and Models*, John Wiley and Sons, Ltd., Chichester, England.

8. Voldman, S. (2007) *Latchup*, John Wiley and Sons, Ltd., Chichester, England.

9. Voldman, S. (2007) *ESD: Design and Synthesis*, John Wiley and Sons, Ltd., Chichester, England.

10. Voldman, S. (2008) *ESD: Circuits and Devices*, Publishing House of Electronic Industry (PHEI), Beijing, China.

11. Ker, M.D. and Hsu, S.F. (2009) *Transient Induced Latchup in CMOS Integrated Circuits*, John Wiley and Sons, Ltd., Singapore.

12. Mardiquan, M. (2009) *Electrostatic Discharge: Understand, Simulate, and Fix ESD Problems*, John Wiley and Sons, Co., New York.

13. Vashchenko, V. and Shibkov, A. (2010) *ESD Design in Analog Circuits*, Springer, New York.

14. Tasca, D.M. (1970) Pulse power failure modes in semiconductors. *IEEE Transactions on Nuclear Science*, **NS-17** (6), 346–372.

15. Wunsch, D.C. and Bell, R.R. (1968) Determination of threshold voltage levels of semiconductor diodes and transistors due to pulsed voltages. *IEEE Transactions on Nuclear Science*, **NS-15** (6), 244–259.

16. Smith, J.S. and Littau, W.R. (1981) Prediction of thin-film resistor burn-out. Proceedings of the Electrical Overstress and Electrostatic Discharge (EOS/ESD) Symposium, pp. 192–197.

17. Ash, M. (1983) Semiconductor junction non-linear failure power thresholds: Wunsch-Bell revisited. Proceedings of the Electrical Overstress and Electrostatic Discharge (EOS/ESD) Symposium, pp. 122–127.

18. Arkihpov, V.I., Astvatsaturyan, E.R., Godovosyn, V.I., and Rudenko, A.I. (1983) Plasma accelerator with closed electron drift. *International Journal of Electronics*, **55**, 135–145.

19. Vlasov, V.A. and Sinkevitch, V.F. (1971) *Elektronnaya Technika*, (4), 68–75.

20. Dwyer, V.M., Franklin, A.J., and Campbell, D.S. (1989) Thermal failure in semiconductor devices. *Solid State Electronics*, vol. 33, 553–560.

21. Brown, W.D. (1972) Semiconductor device degradation by high amplitude current pulses. *IEEE Transactions on Nuclear Science*, **NS-19**, 68–75.

22. Alexander, D.R. and Enlow, E.W. (1981) Predicting lower bounds on failure power distributions of silicon npn transistors. *IEEE Transactions on Nuclear Science*, **NS-28** (6), 4305–4310.

23. Enlow, E.N. (1981) Determining an emitter-based failure threshold density of npn transistors. Proceedings of the Electrical Overstress/Electrostatic Discharge (EOS/ESD) Symposium, pp. 145–150.

24. Pierce, D. and Mason, R. (1982) A probabilistic estimator for bounding transistor emitter-based junction transient-induced failures. Proceedings of the Electrical Overstress/Electrostatic Discharge (EOS/ESD) Symposium, pp. 82–90.

25. Singh, R., Harame, D., and Oprysko, M. (2004) *Silicon Germanium: Technology, Modeling and Design*, John Wiley and Sons.

26. Voldman, S. (1998) The state of the art of electrostatic discharge protection: Physics, technology, circuits, designs, simulation and scaling. Invited Talk, Bipolar/BiCMOS Circuits and Technology Meeting (BCTM) Symposium, pp. 19–31.

27. Voldman, S. (1998) The impact of MOSFET technology evolution and scaling on electrostatic discharge protection. *Journal of Microelectronics Reliability*, **38**, 1649–1668.

28. Voldman, S. (1999) The impact of technology evolution and scaling on electrostatic discharge (ESD) protection in high-pin-count high-performance microprocessors. Proceedings of the International Solid State Circuits Conference (ISSCC), Session 21, WA 21.4, February 1999, pp. 366–367.

29. Voldman, S. (1999) Electrostatic discharge (ESD) protection in silicon-on-insulator (SOI) CMOS technology with aluminium and copper interconnects in advanced microprocessor semiconductor

chips. Proceedings of the Electrical Overstress/Electrostatic Discharge (EOS/ESD) Symposium, pp. 105–115.

30. Voldman, S., Hui, D., Warriner, L. *et al.* (1999) Electrostatic discharge protection in silicon-on-insulator technology. Proceedings of the IEEE International Silicon on Insulator (SOI) Conference, pp. 68–72.

31. Voldman, S. (Feb 25, 2003) Method for evaluating circuit design for ESD electrostatic discharge robustness U.S. Patent No. 6,526,548.

32. ANSI/ESD ESD-STM 5.1 – 2007 (2007) ESD Association Standard Test Method for the Protection of Electrostatic Discharge Sensitive Items - Electrostatic Discharge Sensitivity Testing - Human Body Model (HBM) Testing. Component Level Standard Test Method (STM) document.

33. ANSI/ESD SP 5.1.2-2006 (2006) ESD Association Standard Practice for the Protection of Electrostatic Discharge Sensitive Items - Human Body Model (HBM) and Machine Model (MM) Alternative Test Method: Split Signal Pin-Component Level.

34. Meuse, T., Barrett, R., Bennett, D. *et al.* (2004) Formation and suppression of a newly discovered secondary EOS event in HBM test systems. Proceedings of the Electrical Overstress/Electrostatic Discharge (EOS/ESD) Symposium, pp. 141–145.

35. Ashton, R.A., Weir, B.E., Weiss, G., and Meuse, T. (2004) Voltages before and after HBM stress and their effect on dynamically triggered power supply clamps. Proceedings of the Electrical Overstress/Electrostatic Discharge (EOS/ESD) Symposium, pp. 153–159.

36. Barth, J. and Richner, J. (2005) Voltages before and after current in HBM testers and Real HBM. Proceedings of the Electrical Overstress/Electrostatic Discharge (EOS/ESD) Symposium, pp. 141–151.

37. Gaertner, R., Aburano, R., Brodbeck, T. *et al.* (2005) Partitioned HBM test – A new method to perform HBM test on complex tests. Proceedings of the Electrical Overstress/Electrostatic Discharge (EOS/ESD) Symposium, pp. 178–183.

38. Brodbeck, T. and Gaertner, R. (2005) Experience in HBM ESD testing of high pin count devices. Proceedings of the Electrical Overstress/Electrostatic Discharge (EOS/ESD) Symposium, pp. 184–189.

39. ANSI/ESD ESD-STM 5.2 – 1999 (1999) ESD Association Standard Test Method for the Protection of Electrostatic Discharge Sensitive Items - Electrostatic Discharge Sensitivity Testing - Machine Model (MM) Testing -Component Level. Standard Test Method (STM) document.

40. ANSI/ESD ESD-STM 5.3.1 – 1999 (1999) ESD Association Standard Test Method for the Protection of Electrostatic Discharge Sensitive Items - Electrostatic Discharge Sensitivity Testing – Charged Device Model (CDM) Testing -Component Level. Standard Test Method (STM) document.

41. Voldman, S., Ashton, R., Barth, J. *et al.* (2003) Standardization of the transmission line pulse (TLP) methodology for electrostatic discharge (ESD). Proceedings of the Electrical Overstress/Electrostatic Discharge (EOS/ESD) Symposium, pp. 372–381.

42. ANSI/ESD Association ESD-SP 5.5.1-2004 (2004) ESD Association Standard Practice for the Protection of Electrostatic Discharge Sensitive Items - Electrostatic Discharge Sensitivity Testing – Transmission Line Pulse (TLP) Testing Component Level. Standard Practice (SP) document.

43. ANSI/ESD Association ESD-STM 5.5.1 -2008 (2008) ESD Association Standard Test Method for the Protection of Electrostatic Discharge Sensitive Items - Electrostatic Discharge Sensitivity Testing – Transmission Line Pulse (TLP) Testing Component Level. Standard Test Method (STM) document.

44. ANSI/ESD STM5.5.1-2008 (2008) Electrostatic Discharge Sensitivity Testing – Transmission Line Pulse (TLP) – Component Level.

45. ANSI/ESD STM5.5.2-2007 (2007) Electrostatic Discharge Sensitivity Testing - Very Fast Transmission Line Pulse (VF-TLP) - Component Level.

46. ESD Association ESD-SP 5.5.2 (2007) ESD Association Standard Practice for the Protection of Electrostatic Discharge Sensitive Items - Electrostatic Discharge Sensitivity Testing Very Fast Transmission Line Pulse (VF-TLP) Testing Component Level. Standard Practice (SP) document.

47. ANSI/ESD Association ESD-SP 5.5.2-2007 (2007) ESD Association Standard Practice for the Protection of Electrostatic Discharge Sensitive Items - Electrostatic Discharge Sensitivity Testing – Very Fast Transmission Line Pulse (VF-TLP) Testing Component Level. Standard Practice (SP) document.

48. ESD Association ESD-STM 5.5.2 (2009) ESD Association Standard Test Method for the Protection of Electrostatic Discharge Sensitive Items - Electrostatic Discharge Sensitivity Testing Very Fast Transmission Line Pulse (VF-TLP) Testing Component Level. Standard Test Method (STM) document.

49. Muhonen, K., Ashton, R., Barth, J. *et al.* (2008) VF-TLP round robin study, analysis, and results. Proceedings of the Electrical Overstress/Electrostatic Discharge (EOS/ESD) Symposium, pp. 40–49.

50. ANSI/ESD Association ESD-STM 5.5.1 -2008 (2008) ESD Association Standard Test Method for the Protection of Electrostatic Discharge Sensitive Items - Electrostatic Discharge Sensitivity Testing – Very Fast Transmission Line Pulse (VF-TLP) Testing Component Level. Standard Practice (SP) document.

51. Chen, T.W., Ito, C., Maloney, T. *et al.* (2007) Gate oxide reliability characterization in the 100 ps regime with ultra-fast transmission line pulsing system. Proceedings of the Electrical Overstress/Electrostatic Discharge (EOS/ESD) Symposium, pp. 16–21.

52. Intel Corporation (July 2001) Cable discharge event in local area network environment. White Paper, Order No: 249812-001.

53. Brooks, R. (March 2001) A simple model for the cable discharge event. *IEEE802.3 Cable-Discharge Ad-hoc Committee.*

54. Telecommunications Industry Association (TIA) (December 2002) Category 6 Cabling: Static discharge between LAN cabling and data terminal equipment, *Category 6 Consortium.*

55. Deatherage, J. and Jones, D. (2000) Multiple factors trigger discharge events in Ethernet LANs. *Electronic Design,* **48** (25), 111–116.

56. Stadler, W., Brodbeck, T., Gartner, R., and Gossner, H. (2006) Cable discharges into communication interfaces. Proceedings of the Electrical Overstress/Electrostatic Discharge (EOS/ESD) Symposium, pp. 144–151.

57. ESD Association DSP 14.1-2003 (2003) ESD Association Standard Practice for the Protection of Electrostatic Discharge Sensitive Items – System Level Electrostatic Discharge Simulator Verification Standard Practice. Standard Practice (SP) document.

58. ESD Association DSP 14.3-2006 (2006) ESD Association Standard Practice for the Protection of Electrostatic Discharge Sensitive Items – System Level Cable Discharge Measurements Standard Practice. Standard Practice (SP) document.

59. ESD Association DSP 14.4-2007 (2007) ESD Association Standard Practice for the Protection of Electrostatic Discharge Sensitive Items – System Level Cable Discharge Test Standard Practice. Standard Practice (SP) document.

60. Geski, H. (September 2004) DVI compliant ESD protection to IEC 61000-4-2 level 4 standard. *Conformity,* pp. 12–17.

61. International Electro-technical Commission (IEC) IEC 61000-4-2 (2001) Electromagnetic Compatibility (EMC): Testing and Measurement Techniques – Electrostatic Discharge Immunity Test.

62. Grund, E., Muhonen, K., and Peachey, N. (2008) Delivering IEC 61000-4-2 current pulses through transmission lines at 100 and 330 ohm system impedances. Proceedings of the Electrical Overstress/Electrostatic Discharge (EOS/ESD) Symposium, pp. 132–141.

63. IEC 61000-4-2 (2008) Electromagnetic Compatibility (EMC) – Part 4-2: Testing and Measurement Techniques – Electrostatic Discharge Immunity Test.

64. Chundru, R., Pommerenke, D., Wang, K. *et al.* (2004) Characterization of human metal ESD reference discharge event and correlation of generator parameters to failure levels – Part I: Reference Event. *IEEE Transactions on Electromagnetic Compatibility*, **46** (4), 498–504.

65. Wang, K., Pommerenke, D., Chundru, R. *et al.* (2004) Characterization of human metal ESD reference discharge event and correlation of generator parameters to failure levels – Part II: Correlation of generator parameters to failure levels. *IEEE Transactions on Electromagnetic Compatibility*, **46** (4), 505–511.

66. ESD Association ESD-SP 5.6 -2008 (2008) ESD Association Standard Practice for the Protection of Electrostatic Discharge Sensitive Items - Electrostatic Discharge Sensitivity Testing – Human Metal Model (HMM) Testing Component Level. Standard Practice (SP) document.

67. ANSI/ESD SP5.6-2009 (2009) Electrostatic Discharge Sensitivity Testing - Human Metal Model (HMM) - Component Level.

68. Antognetti, P. (1986) *Power Integrated Circuits: Physics, Design, and Applications*, McGraw-Hill, New York.

69. Hower, P.L. and Govil, P.K. (1974) Comparison of one- and two-dimensional models of transistor thermal instability. *IEEE Transactions of Electron Devices*, **ED-21** (10), 617–623.

70. Hower, P., Lin, J., Haynie, S. *et al.* (1999) Safe operating area considerations in LDMOS transistor. International Symposium on Power Semiconductors and IC s (ISPSD), pp. 55–84.

71. Hower, P. and Pendeharker, S. (2005) Short and long-term safe operating area considerations in LDMOS transistors. Proceedings of the International Reliability Physics Symposium (IRPS), pp. 545–550.

72. Liou, J.J., Malobabic, S., Ellis, D.F. *et al.* (2009) Transient safe operating area (TSOA) definition for ESD applications. Proceedings of the Electrical Overstress/Electrostatic Discharge (EOS/ESD) Symposium, pp. 17–27.

73. Gray, P., Hurst, P., Lewis, S. and Meyer, R. (2009) *Analysis and Design of Analog Integrated Circuits*, 5th edn, John Wiley and Sons, Co., New York.

74. Sansen, W.M.C. (2006) *Analog Design Essentials*, Springer, Netherlands.

75. Hastings, A. (2006) *The Art of Analog Layout*, 2nd edn, Pearson Prentice Hall, New Jersey.

76. Ghandi, S.K. (1977) *Semiconductor Power Devices*, John Wiley & Sons, New York.

77. Antognetti, P. (1986) *Power Integrated Circuits: Physics, Design, and Applications*, McGraw-Hill, New York.

78. Baliga, B.J. (1988) *High Voltage Integrated Circuits*, IEEE Press, New York, N.Y.

79. Baliga, B.J. (1987) *Modern Power Devices*, John Wiley and Sons, Inc., New York, N.Y.

80. Vashchenko, V., Ter Beek, M., Kindt, W., and Hopper, P. (2004) ESD protection of high voltage tolerant pins in low voltage BiCMOS processes. Proceedings of the Bipolar Circuits Technology Meeting (BCTM), pp. 277–280.

81. Jowett, C.E. (1976) *Electrostatics in the Electronic Environment*, Halsted Press, New York.

82. Lewis, W.H. (1995) *Handbook on Electromagnetic Compatibility*, Academic Press, New York.

83. Morrison, R. and Lewis, W.H. (1990) *Grounding and Shielding in Facilities*, John Wiley and Sons Inc., New York.

84. Paul, C.R. (2006) *Introduction to Electromagnetic Compatibility*, John Wiley and Sons Inc., New York.

85. Morrison, R. and Lewis, W.II. (2007) *Grounding and Shielding*, John Wiley and Sons Inc., New York.

86. Ott, H.W. (2009) *Electromagnetic Compatibility Engineering*, John Wiley and Sons Inc., Hoboken, New Jersey.

87. Ott, H.W. (1985) Controlling EMI by proper printed wiring board layout. Sixth Symposium on EMC, Zurich, Switzerland.

88. ANSI C63.4-1992 (July 17 1992) *Methods of Measurement of Radio-Noise Emissions from Low-Voltage Electrical and Electronic Equipment in the Range of 9 kHz to 40 GHz*, IEEE.

89. EN 61000-3-2 (2006) *Electromagnetic Compatibility (EMC) – Part 3-2: Limits-Limits for Harmonic Current Emissions (Equipment Input Current < 16 A Per Phase)*, CENELEC.

90. EN 61000-3-3 (2006) *Electromagnetic Compatibility (EMC) – Part 3-3: Limits-Limitation of Voltage Changes, Voltage Fluctuations and Flicker in Public Low-Voltage Supply Systems for Equipment with Rated Current < 16A Per Phase and Not Subject to Conditional Connection*, CENELEC.

91. EN 61000-4-2 (2001) Electromagnetic Compatibility (EMC) – Part 4-2: Testing and Measurement Techniques – Electrostatic Discharge Immunity Test.

92. MDS MDS-201-0004 (October 1 1979) *Electromagnetic Compatibility Standards for Medical Devices*, U.S. Department of Health Education and Welfare, Food and Drug Administration.

93. MIL-STD-461E (August 20 1999) Requirements for the Control of Electromagnetic Interference Characteristics of Subsystems and Equipment.

94. RTCA RTCA/DO-160E (December 7 2004) *Environmental Conditions and Test Procedures for Airborne Equipment*, Radio Technical Commission for Aeronautics (RTCA).

95. SAE SAE J551 (June 1996) *Performance Levels and Methods of Measurement of Electromagnetic Compatibility of Vehicles and Devices (60 Hz to 18 GHz)*, Society of Automotive Engineers.

96. SAE SAE J1113 (June 1995) *Electromagnetic Compatibility Measurement Procedure for Vehicle Component (Except Aircraft) (60 Hz to 18 GHz)*, Society of Automotive Engineers.

97. Wall, A. (2004) Historical Perspective of the FCC Rules for Digital Devices and a Look to the Future. IEEE International Symposium on Electromagnetic Compatibility, August 9–13, 2004.

98. Denny, H.W. (1983) *Grounding For the Control of EMI*, Don White Consultants, Gainesville, VA.

99. Boxleitner, W. (1989) *Electrostatic Discharge and Electronic Equipment*, IEEE Press, New York.

100. Gerke, D.D. and Kimmel, W.D. (March/April 1986) Designing noise tolerance into microprocessor systems. *EMC Technology*.

101. Kimmel, W.D. and Gerke, D.D. (September 1993) Three keys to ESD system design. *EMC Test and Design*.

102. Violette, J.L.N. (May/June 1986) ESD case history – Immunizing a desktop business machine. *EMC Technology*.

103. Wong, S.W. (May/June 1984) ESD design maturity test for a desktop digital system. *Evaluation Engineering*.

104. Voldman, S., Never, J., Holmes, S., and Adkisson, J. (1996) Linewidth control effects on MOSFET ESD robustness. Proceedings of the Electrical Overstress/Electrostatic Discharge (EOS/ESD) Symposium, pp. 101–109.

4 System Level ESD

A summer student working in a server division plugged a cable into a server without discharging the cable to the chassis of the system. The pulse from the cable entered the signal pin of the first semiconductor component. The cable discharge event was a negative polarity pulse. The negative pulse was discharged through an n-well to substrate ESD diode within the semiconductor chip. The circuit designers and ESD engineer forgot to place a guard ring around the n-well diode. The guard ring of the I/O standard cell circuit was too small and too resistive to sink the high current from the cable. The cable discharge pulse propagated into the adjacent low voltage CMOS circuitry, leading to a latchup event. The latchup event propagated across the semiconductor chip, leading to over-current, and melting of the semiconductor chip package. Smoke came out of the server, and the fire detectors in the building went off, turning on the water from the fire prevention system. Fire trucks from the local fire department also visited to help out.

4.1 SYSTEM LEVEL TESTING

System level testing is very important in the qualification of systems to evaluate the electromagnetic compatibility (EMC), electromagnetic emissions (EMI) sensitivity, as well as vulnerability to different sources of electrical overstress (EOS). In this chapter, system level concerns associated with ESD, EOS, latchup, EMI and EMC will be discussed relevant to today's application and the future.

A brief discussion of electrostatic issues in servers, laptops, hand-held devices, cell phones, disk drives, digital cameras, autos, and space applications is highlighted to educate the reader in awareness of the vast number of issues in today's electronic environment.

In the past, system level testing was very distinct from the component level testing; today, in some areas there is a convergence of the component and system level tests. In this chapter

ESD Basics: From Semiconductor Manufacturing to Product Use, First Edition. Steven H. Voldman.
© 2012 John Wiley & Sons, Ltd. Published 2012 by John Wiley & Sons, Ltd.

we will discuss the distinction between the system level tests as well as the convergence (or overlap) of the system level and component level testing. The chapter will discuss ESD [1–9], and latchup [7,8] with respect to their issues on systems. System level and "system level-like" ESD specifications, such as IEC 61000-4-2 [10–13], human metal model (HMM) [14–17], and cable discharge event [18–25] will be discussed relevant to products and systems. EMC and EMI will be discussed relevant to today's product applications.

4.1.1 System Level Testing Objectives

System level testing has fundamental performance objectives and criteria. Product manufacturers must meet the following requirements:

* No degradation or loss of function during the system test.
* No degradation below a specified level by the manufacturers.
* Temporary loss of function allowed provided self-recovery or operator recover is possible.
* No malfunction or upset during installation and/or assembly.

System level susceptibility includes the loss of operation of the system during the applied tests. In system level testing, it is not necessary to have destructive failure or permanent degradation of the components of the system. A first key point is that system level testing is to insure the system functions during the applied test. A second point is that during system level testing, the system is in a powered state (e.g. it is "turned on").

4.1.2 Distinction of System and Component Level Testing Failure Criteria

The system level susceptibility is very distinct from the component level failure criteria.

One of the reasons is the objectives are very different. In this section, we will highlight the important distinctions (Figure 4.1).

* Powered vs. Unpowered.
* Destructive vs. Non-destructive Testing.

In the testing of systems, the system is powered during the test, whereas in semiconductor component testing, the components are unpowered. The semiconductor components are tested as socketed or un-socketed, depending on the test.

In the testing of systems, the system is evaluated *during* the test, as well as after. The testing of the systems by system manufacturers must verify that the devices or components function *during* the test itself. ESD testing of components by semiconductor manufacturers insure devices will function *after* an ESD event that might occur during handling.

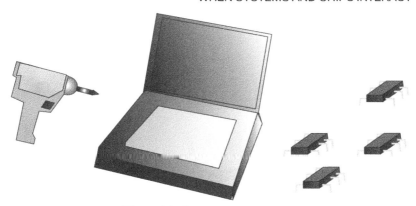

Figure 4.1 Systems vs. components

Semiconductor device level testing is tested to destructive failure intentionally to understand the capability of the component to survive the handling process prior to delivery to the system manufacturer. The device handling procedures is associated with the semiconductor component level ESD test, such as the human body model (HBM), machine model (MM), and charged device model (CDM). Semiconductor device level testing at times will test to a given specified degradation level of the component, where the specified degradation level is affiliated with a semiconductor chip electrical criteria. For example, transistor electrical parameters can vary with ESD testing below the manufacturers' specified electrical parameters.

From this discussion, it is clear that the objectives of the two are not equivalent.

4.2 WHEN SYSTEMS AND CHIPS INTERACT

Systems and components are interactive in the testing of systems, and yet the system level objectives are not aligned with the component level objectives.

A first dilemma between system level performance and component level evaluation exists because of the distinct and different objectives. Component level tests are rated to survive ESD in the handling process, but a chip's survivability during ESD testing is not a measure of its susceptibility level in the system (Figure 4.2).

A second dilemma is that system level performance and susceptibility is impacted by component susceptibility. If they were totally independent then this would not be an issue; but, it is the components that lead to system level failure.

This poses the following questions:

- How do you quantify a system susceptibility when the component distributors test only the devices to ESD tests?

- How do you bridge from semiconductor ESD robustness to system level susceptibility?

- How do you bridge from semiconductor ESD robustness to system level susceptibility when the failure criteria are different from components versus systems?

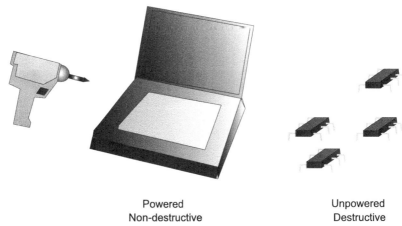

Powered   Unpowered
Non-destructive   Destructive

Figure 4.2   Component level upset versus system level upset

Today, there is a trend to apply "system-like" tests to semiconductor chip components. These include the following tests:

- Cable discharge event (CDE).

- Charged board event (CBE).

- IEC 61000-4-2 system pulse.

- Human metal model (HMM).

These "system" and "system-like" tests will be discussed in later sections.

4.3 ESD AND SYSTEM LEVEL FAILURES

In the ESD testing of components, the criteria of failure is when the component undergoes damage or physical change. The physical change is typically associated with leakage or parameter changes. These ESD failures typically occur in the signal pin circuitry (e.g. receivers and off-chip drivers), or ESD networks. But, for system level failures, the ESD current itself (e.g. non-destructive) can "upset" or interrupt system operation. Figure 4.3 shows different components in different packages.

4.3.1 ESD Current and System Level Failures

The ESD current can be applied as follows:

- Ground Injection: Injection directly into the ground through a shielded enclosure, grounded connector shell.

Figure 4.3 Packaged components

- Ground Injection via ESD Protection Networks: Shunted to the ground through on- or off-chip ESD protective devices via a defined ESD current path.

- Signal Path Injection: Injection directly into the signal path or an unprotected circuit.

In the first case, the current is injected into the ground through a connector, or shielded enclosure. The ESD current propagates into the system between the shield and the ground connections.

In the second case, the ESD current is injected to the ground through on-chip ESD protection circuitry, or diverted by off-chip ESD protection circuitry. On-chip protection circuitry is integrated with the semiconductor chip bond pads, and circuits. Off-chip protection are mounted on the card or board connected to the signal pins and ground.

In the third case, the current injection goes directly into the signal path and the current flows through the signal path circuitry back to the ground plane.

4.3.2 ESD Induced E- and H-Fields and System Level Failures

The ESD current itself flowing through the alternative current loop established by the protection circuitry can be a source of system level failures. The ESD current flowing through the component or semiconductor chip, board or other system components can lead to a system upset. The ESD current will induce electric field (E) and magnetic field (H) in the system (Figure 4.4). The E- and H-fields are created, radiated, propagated and are absorbed in the boards, packaging, and components. These ESD-induced E- and H-fields can cause semiconductor devices malfunction. In some cases, the ESD currents into unprotected devices are more likely to cause hard failures than upset.

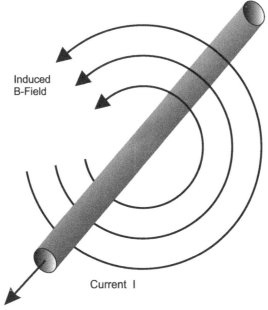

Induced
B-Field

Current I

Figure 4.4 Induced H-field

4.4 ELECTRONIC SYSTEMS

Electronic systems contain discrete electrical components, packaged semiconductor integrated circuit chips, cards, boards, chassis, and exterior assembly. Each element within the electronic system can be impacted by EOS, ESD, and EMI. Electromagnetic compatibility (EMC) also is a concern in electronic assemblies.

4.4.1 Cards and Boards

Discrete elements, and integrated semiconductor chips are mounted on cards and boards in electronic systems. These components have electrical resistance, inductance, and capacitance. When the components are not mounted on a system level assembly, they are vulnerable to current sources, voltage sources, and static charging.

In the case of charging, the component has an internal capacitance, but also a capacitance to the ground plane (or table surface). Given the component is within an external electric field, the element can charge up given there is a connection to the component. The component can also charge when electrically connected to a voltage source. In the "charged device model" the component is first charged by an external connection to a voltage source. The voltage source is removed, and then the charged component maintains its charge. The larger the component, the more the capacitance, and the higher the total charge stored. When an external ground is applied to any of the signal pins, the charge from within the semi-conductor component flows to the grounded signal pin.

With the introduction of a card, or board, the combination of the components and the card is now a larger capacitance to the ground plane. Given the same charging procedure is applied, more total charge is now stored on the card-chip assembly. In this case, a trace on the card is grounded and the current flows through the complete assembly.

4.4.2 System Chassis and Shielding

In system design, a system "chassis" or shield is used to avoid electrical emissions, and noise interference. System failures can occur from the ESD current itself (e.g. non-destructive) which can "upset" or interrupt system operation. The ESD current can be applied directly into the ground through a shielded enclosure, grounded connector shell, or shielding.

In a non-contact electrical event, the discharge current can produce an electrical arc. The electrical arc from the source to the applied system produces electrical current, as well as an E-field and H-field. Some of the energy is in the direct current, and some of the energy is in the electromagnetic field. The system chassis and shielding is intended to block out the elec-tromagnetic field outside of the electrical system.

4.5 SYSTEM LEVEL PROBLEMS TODAY

Today, with the ever-changing systems, the ESD, EOS, EMI and EMC issues change. Prior to in-depth discussion, a brief discussion on different "systems" will be held to make the issues relevant to today's world.

4.5.1 Hand Held Systems

Today, the number of systems that are hand held has increased significantly from a few gen-erations ago. Hand-held devices extend from electronic games, USB memory sticks, cell phones, iShuffles, iPods, to iPads. (Figure 4.5) The landscape of systems has changed from large "big iron" computers that occupy complete floors to small electronic components. In the early days, the interaction with the large system was addressed with computer operators and control procedures. In the past, the only electronic systems were electronic calculators. The world of handheld electronics went from the "Palm" and electronic games to a plethora of electronic devices.

The human interaction with these components has also changed from keyboards to new "touch" technology. In the early days, there was interaction with buttons and keyboards. Today, the human interaction is with touch screens, and the electrical connection ports.

4.5.2 Cell Phones

In the early development of cell phones in the mid-1990s the first cell phones were large and cumbersome. The cell phones had keys to push similar to a standard desk top telephone. The

Figure 4.5 Handheld devices

keys on the cell phones were possible exposure points for electromagnetic interference (EMI), to electrostatic discharge (ESD). The number of cell phones at that time was small relative to the magnitude of cell phones in the world today.

Today, and in the future, cell phones have quickly migrated to very small systems that can be placed in a shirt pocket, pant pocket, or purse. Today's cell phones are either standard cell phones or smart phones.

Today's smaller cell phones still have keyboards, keys and ports where ESD can access the cell phones. One of the present problems is ESD events that occur on the exposed ports of the cell phone. The cell phone antennae are electrically connected to a gallium arsenide power amplifier that serves as a transmitter of the signal. Electromagnetic signals and ESD events can occur on the port where the antennae exist. The current can enter the cell phones, and lead to failure of the GaAs bipolar transistor that is contained within the cell phone through the ports. Figures 4.6 and 4.7 show ESD failures in a semiconductor component within a cell phone [5].

Today, many cell phones have migrated to smart phone technology. With smart phones, the keys and keyboard are now "touch" technology. With "touch" technology, there are no openings in the keys, and all human interaction is done through the touch screen. The ports for the smart phones are still exposed, but significantly fewer. The connectors to computers and power are designed to avoid electrical connection to signal pins prior to the connection of the ground and power pins.

In the future, it is anticipated that "touch-less technology" will replace touch technology. In that case, the direct human interaction will reduce the risk of charging phenomena.

4.5.3 Servers and Cables

It was not long ago that big systems were connected to "dummy terminals" and centralized computing. In those days, the large systems had handling procedures for the assembly of the

Gallium Arsenide

DCS Mode : 6um 900 ohm resistor

Figure 4.6 ESD failure in a cell phone containing a GaAs semiconductor chip

systems. After the machines were installed, electrical procedures were present for connecting or disconnecting cables. It was common practice to have "touch pads" on the large computers to discharge the cable and the operator prior to insertion of the cable to the system. The procedure was for the operator to hold the cable, and then discharge the cable and himself to the "touch pad". The "touch pad" allowed the ESD current to discharge to the system ground avoiding ESD current from entering the system. Operators were trained to follow these procedures due to the cost of the computers at times exceeding 10 million dollars for each large system.

Gallium Arsenide

GSM Mode: 6um 2K ohm resistor

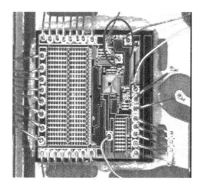

Figure 4.7 ESD failure in a cell phone containing a GaAs semiconductor chip. Permission granted from ESD Association

Figure 4.8 Laptops and cables

4.5.4 Laptops and Cables

As large system environments moved to small servers and the creation of desk tops, laptops and other small computing systems, it was no longer possible to have controls for cable insertions and system-level procedures. Today, laptops have ports that are exposed and this is where ESD failure can occur.

In present day laptops, ESD, EOS, and EMI can enter the system through the screen, the keyboards, and ports. Testing of laptop computers is initiated by applying the IEC 61000-4-2 system test to all locations on the laptop computer. Laptop ports are sources of potential concern where USB and cables are connected. (Figure 4.8)

One source of potential charging is the placement of the USB element. When a USB element is installed into a computer, the charge from the person and the USB stick itself can be charged prior to insertion into the computer. Today, USB sticks are transferred from system to system without thought about the charging and discharging process.

4.5.5 Disk Drives

In the disk drive industry, the magnetic recording heads are vulnerable to electrostatic discharge (ESD), electromagnetic interference (EMI), as well as issues with electromagnetic compatibility (EOS) [5] (Figure 4.9).

Magneto-resistor technology is used for sensing the signal on the disk. MR technology does not have ESD solutions integrated with the magnetic recording heads for both the "read" and "write" operations.

In the manufacturing line, MR heads are sensitive to EMI noise in the manufacturing environment. MR head failures can occur when the electromagnetic field induces a voltage exceeding the voltage-to-failure of the magnetic resistor stripe. The MR head failures will

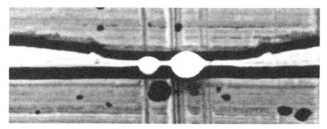

Figure 4.9   ESD failure in disk drives

occur from induced voltages in the order of 2V for an MR head, 1V for a giant magneto-resistor (GMR) head, and 0.5 V for a tunneling magneto-resistor (TMR) head.

Due to the ESD sensitivity, washing operations, and shipping can also lead to MR head failures. ESD failures occur due to charging which can lead to MR head failure mechanisms. ESD failure mechanisms include MR heads, MR stripe, MR stripe-to-shield breakdown, inductor write head failure, inductor-to-yoke breakdown, and write head-to-substrate break-down [5].

ESD related failures also lead to changes in the magnetic characteristics of the MR head. The ESD current leads to self-heating of the MR stripe, causing changes in the relationships of the induced electrical differential voltage vs magnetic field response relationship. ESD failures also induce resistance changes in the MR stripe, leading to functional failures.

Magnetic recording heads are mounted onto an armature structure when it is in operation. When the magnetic head is in the armature, EOS and ESD events on the armature can propa-gate through the armature assembly. The pulse can propagate through the pre-amplifier chip, through small wire connections and to the MR head. In this process, the pulse waveform is modified, but can still lead to ESD failures.

ESD changes also influence the operation of the disk drive after assembly. Since the MR stripe must fly across the disk, even the changes in the MR stripe geometry affect the aero-dynamic flight characteristics of the magnetic head.

4.5.6 Digital Cameras

Early cameras consisted of the Kodak box camera (Figure 4.10). Figure 4.10 shows the Kodak "Brownie" camera. The earliest cameras had a roll of film on a spool, and a mechani-cal winding knob to rotate the spool. The shutter was a mechanical shutter. The roll of film was put in the camera by Kodak. The person would "shoot" his pictures, and send the camera back to Kodak, where they would develop the pictures, and re-load a new roll of film.

As cameras advanced, the magnitude of the electronics increased. Advancements included electronic shutters and motor drives to wind the film. Single lens reflex (SLR) cam-eras slowly added more electronics for the photographer but still contained a film to be proc-essed. With the introduction of the digital camera, came cameras with few mechanical features and a very significant amount of electronics. Digital cameras use CMOS image processing semiconductor chips to capture the image, and store it on a memory chip.

Figure 4.10 Early Kodak camera

The sequence of the powering of the camera and the memory card is important to avoid electrical impact to the camera or the memory card circuitry. Warnings are present in the camera literature:

- Turn the power off before inserting or removing memory cards.

- Do not remove the memory card from the camera, turn the camera off, or remove or disconnect the power source during formatting or while data is being recorded, deleted, or copied to a computer.

- Failure to observe these precautions could result in loss of data or in damage to the camera or card.

- Do not touch the card terminals with your finger or metal objects.

The images stored on the camera can be transferred to a printer, or computer. On my digital SLR camera, there are three ports for data transfer: an accessory terminal, USB connector, and a HDMI mini-pin connector. With the integration of digital cameras, the computers, and printers, the sequence of the turn-on of the camera, and computer is important for the I/O circuitry and discharge of residual charge on the cables. An example sequence is as follows:

- Turn the camera off.
- Turn the computer on.
- Connect the USB cable.
- Turn the camera on.
- Transfer photographs.
- Turn the camera off and disconnect the USB cable.

  Warnings are present in the camera literature:

- Be sure the camera is off when connecting and disconnecting interface cables.
- Do not turn the camera off or disconnect the USB cable while transfer is in progress.

  With the emergence of smart phone cameras, camera concerns will lessen. In smart phones the memory is embedded into the smart phones. The cabling between the smart phone and the computer is sequence independent. Hence, today's smart phones have eliminated some of the memory insertion and handling issues, sequencing issues, and exposure to electrostatic discharge.

4.6 AUTOMOBILES, ESD, EOS, AND EMI

Electrostatic discharge (ESD), electrical overstress (EOS), and electromagnetic interference (EMI) issues are a significant concern in the auto industry. Key areas of interest today in automobiles are ignition system, electronic pedal systems, gas tank fires, to high voltage hybrid and electric cars.

4.6.1 Automobiles and ESD – Ignition Systems

Automobile ignition systems are ESD sensitive. When a person gets into his automobile, the person can be tribo-electrically charged. Given that the material of the seating is not static dissipative, when the operator sticks his key into the ignition system, a high current spark can enter the ignition electronics. This concern is more significant in cold dry locations, where tribo-electrical charging can occur. The ESD pulse goes into the ignition electronics and can lead to malfunction, or failure of the ignition system.

4.6.2 Automobiles and EMI – Electronic Pedal Assemblies

In automobiles, the gas pedal was a mechanical system whose operation was a function of the operator's pressure on the pedal. Over time, these systems became electrical systems.

One concern in these systems is that malfunction can occur from electromagnetic interference (EMI) noise from other sub-systems under the hood of the automobile. EMI noise isolation is important between the sub-systems of an automobile. As cars continue to add more electronics under the hood, EMI and EMC will be a growing issue.

4.6.3 Automobiles and Gas Tank Fires

In automobiles, gas tank fires are a concern which involves an electrostatic discharge. In cold dry environments, significant charging leading to gas tank fires can occur. When the gasoline is pumped into a tank, there are ions in the gasoline. The charged ions flow into the automobile gas tank. The charged ions in the gasoline migrate to the wall of the gas tank, charging the automobile. Since the automobile is on rubber tires, the insulating nature of the rubber tires allows the automobile to continue to charge.

In cold environments, the driver returns to the automobile, and sits inside it since he or she does not want to stand outside in the cold weather. In the meantime, as the gas pumps into the tank, the gaseous fumes flow out as the gas flows in. When the driver removes the gas pump handle from the gas tank, an electrical spark can occur between the automobile and the gas metal handle. The electrical spark can ignite the gaseous fumes flowing out of the tank, leading to a fire.

Today, when you go to a gas station, there are electrostatic warnings on the gas pump (Figure 4.11). To prevent the problem from occurring, one can stand outside the automobile during the pumping process. Second, if you place your hand on the automobile, you will form a resistance path back to the earth ground plane, dissipating the charge of the automobile. Third, you can also ground yourself to the automobile, prior to handling of the gas pump handle. Figure 4.11 is a warning for filling up a gas can, and an incorrect, and correct procedure.

4.6.4 Hybrids and Electric Cars

With the introduction of hybrids and electric cars, there is even more growth of high voltage electronics inside of automobiles. In hybrids and electric cars, the electrical voltages have increased. In these electronic systems, the ESD robustness of the discrete components or integrated electronics is greater. Increasing electrical power in vehicles introduced powerful high-voltage systems consisting of high-voltage components, related wiring harnesses, and dedicated safety concepts.

In a hybrid automobile, the high voltage system is isolated from other sub-systems. The high voltage system location is isolated, so too are the battery compartments. Fuses isolate the high voltage system. High voltage cables are color coded as orange. The introduction of high voltage systems must take into account:

- Legislative requirements with respect to EMC/EMI and safety such as EN 61508.

- Specifications (ISO, DIN, SAE and GB).

- Special technical problems such as "arcing".

Figure 4.11 Gas pump ESD warnings

In these high voltage systems, the design of the high voltage power supplies, DC/DC converters, relays, fuses, shielded/high voltage cables, insulation, and wiring harnesses are part of the high voltage sub-system.

In an emergency, removal of the ignition key and disconnecting a vehicle's 12-volt battery are common first-responder tasks. Performing that task on a hybrid disables its high-voltage controller. In addition, extinguishing a hybrid vehicle fire can be addressed with large amounts of water, which will both eliminate the radiant heat and also cool the hybrid's metal battery box and the plastic cells inside the battery pack.

4.6.5 Automobiles in the Future

In the future, with hybrid and electric cars, the growth of the electronics in automobiles will only increase. Hence ESD, EOS, EMI and EMC will be important in automobile sub-systems and electrical isolation between all the sub-systems. In the near future, 77 GHz automobile radar for collision protection and initiation of safety features such as airbag initiation will be introduced. Electrical substations will exist as a power source for charging batteries. It is anticipated that to meet current safety standards, significant work will be needed in the auto industry.

4.7 AEROSPACE APPLICATIONS

In the aerospace industry, ESD, EOS, latchup, EMI and EMC are important for airplanes, spacecraft, and satellites. In the following sections, some of the issues in this vast area are highlighted.

4.7.1 Airplanes, Partial Discharge, and Lightning

Airplanes are concerned with electrical overstress (EOS) from lightning and "partial discharge." Lightning that we observe from the ground is typically a discharge between the clouds and the earth. There is a second type, which occurs between two clouds of different electrostatic potential, known as partial discharge. Airplanes flying between the clouds must be concerned with both phenomena.

When I was a graduate student at the MIT High Voltage Research Laboratory, MIT faculty members, such as MIT Professor Cook, studied the phenomena of partial discharge. The simulation experiments would include the following process:

- *An insulator block was placed under a high energy electron generator.*
- *A high energy electron beam would inject electrons into an insulator material.*
- *The electron beam was turned off, and the insulator was removed.*
- *In the insulator, dendritic patterns were observable, and flashing photon discharges could be visibly observed.*
- *A metal object, such as a nail, was then placed at the location of the beam entry point on the insulator.*
- *The metal object has an electrical ground wire.*
- *A hammer was used to pound the nail into the insulator, and then as the discharge occurred, partial discharges and light was observable.*

The nail represented a metallic object penetrating through the insulator region, altering the electric field, and causing additional partial discharges.

Figure 4.12 is the insulator block highlighting the discharge "tree" pattern embedded within the insulator.

Figure 4.12 Partial discharge experiment

4.7.2 Satellites, Spacecraft Charging, and Single Event Upset (SEU)

Satellites can exhibit problems due to electrostatic charging, to single event upset (SEU) events. Electrostatic charging can be due to charged ions that are accumulated on solar panels and large areas on satellites. Due to the environment, the charge accumulated does not have dissipative surfaces in order to reduce the cumulative effect of the charging. When the accumulated charge achieves electrical breakdown, the current can enter the electronics contained within the satellite element. Semiconductor components in the system that are not robust may lead to failure of satellite sub-systems. Satellite components that are typically the weakest are Gallium Arsenide components.

High energy particles, such as heavy ions, can penetrate into the satellite leading to component failures. Single event upsets (SEU) can lead to failure of the electronic component, to soft errors. SEUs that cause permanent damage can include Single Event Gate Ruptures (SEGR). Soft errors lead to flipping of the memory states within the memory chips. Single event particles can also lead to Single Event Latchup (SEL) issues (Figure 4.13).

4.7.3 Space Landing Missions

As is true for satellites, space missions are vulnerable to charging phenomena, cosmic rays to heavy ion events. Landing vehicles, rovers, and space station systems can fail due to electrostatic charging effects on solar panels. Single event latchup (SEL) can also occur within the integrated semiconductor components from heavy ion particles. Semiconductor components, such as CMOS technology are vulnerable to SEL phenomena. Silicon on Insulator (SOI) technology is utilized in space applications to avoid parasitic thyristors, and avoid charging within the semiconductor chip substrate wafer. Figure 4.14 shows an SOI device with a single event. In SOI, a large number of the events occur below the buried oxide (BOX) film, preventing the deep minority carriers from influencing the devices at the surface.

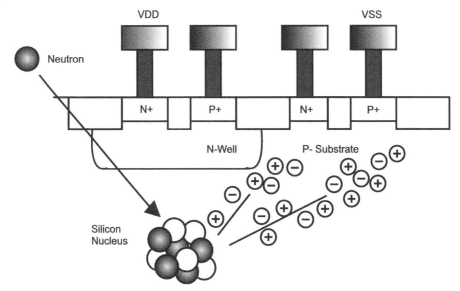

Figure 4.13   Single event latchup (SEL)

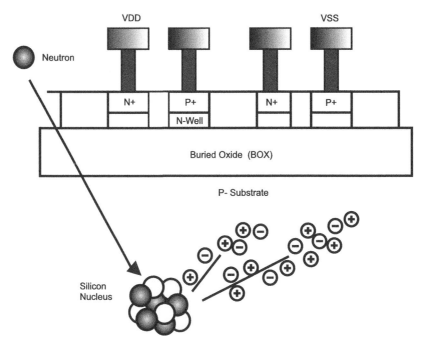

Figure 4.14   Silicon on insulator (SOI) device

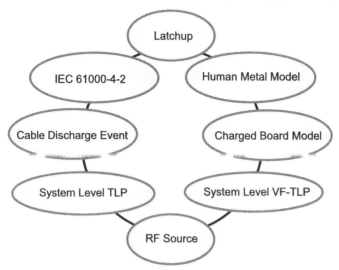

Figure 4.15 System and system-like ESD events

4.8 ESD AND SYSTEM LEVEL TEST MODELS

In the following sections, system level, and "system-like" ESD events will be discussed. The ESD tests of interest are the IEC 61000-4-2 test [10–13], the human metal model (HMM) test [14–17], the charged board model (CBM), cable discharge event [18–25], and latchup [7,8]. This is followed by discussions on electromagnetic compatibility (EMC) testing of components and systems [26–48]. Figure 4.15 shows the type of tests performed on systems. Although not a standard for systems, TLP, VF-TLP, and RF sources are used on systems as well for evaluation of the power-to-failure or robustness of systems from incoming pulse and continuous events.

4.9 IEC 61000-4-2

For system level testing, an ESD gun is used as a pulse source. This IEC system level test standard uses an ESD gun which provides an arc discharge from the gun to the system under investigation [10–13]. In system level testing for ESD, system level designers are interested in both the ESD current discharge and the electromagnetic emissions (EMI) produced by the arc discharge process. In a system, the electromagnetic emissions also can impact the electronics or components. In a system environment, the metal casing around the electronics is to form a Faraday cage, and avoid penetration of the EMI into the electronics. Figure 4.16 is the IEC test configuration for applying the ESD gun pulse to the system under test. Figure 4.17 shows the IEC 610004-2 waveform for the IEC test.

4.10 HUMAN METAL MODEL (HMM)

In the past, ESD testing was performed on semiconductor components. Today, there is more interest in the testing of components in powered states, and in electrical systems. System

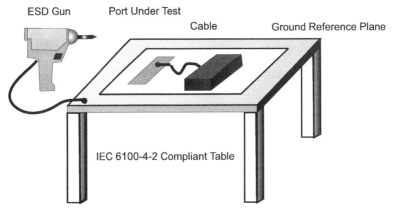

Figure 4.16 IEC test configuration

manufacturers have started to require system level testing to be done on semiconductor components, prior to final assembly and product acceptance. These system level tests are performed with an ESD gun, and without direct contact; these air discharge events produce both an ESD event as well as generating EMI. In a true system, the system itself provides shielding from EMI emissions. Hence, an ESD test is of interest if it has the following characteristics [14–17]:

- An IEC 61000-4-2 current waveform.

- No air discharge (contact discharge).

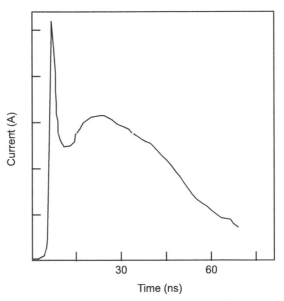

Figure 4.17 IEC 610004-2 current waveform

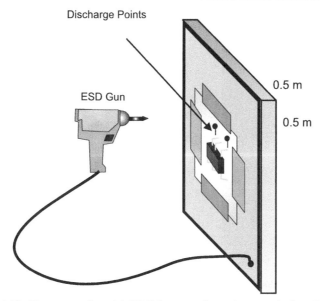

Figure 4.18 Human metal model (HMM) test configuration – vertical configuration

- Semiconductor component is powered during ESD testing.
- Only addresses pins and ports exposed to the external system.

The human metal model (HMM) addresses these characteristics [14–17]. The HMM event is a recent ESD model which has increased interest as a result of cell phone and small components with exposed ports, where field failures were evident. The HMM uses IEC-like pulse waveforms. The discharge from the source and the DUT is a direct contact to avoid EMI spurious signals. The test is performed when the system is powered, and only the external ports that are exposed to the outside world are of interest. Figures 4.18 and 4.19 show the test system configurations, where the source is an ESD gun.

For the human metal model (HMM) test, the current waveform probe must be verified to demonstrate compliance with the HMM specification (Figure 4.20). Figure 4.21 shows a HMM waveform verification.

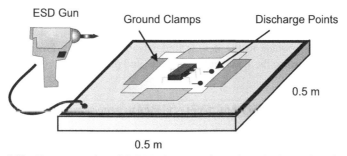

Figure 4.19 Human metal model (HMM) test configuration – horizontal configuration

Probe Verification Methodology

Figure 4.20   Human metal model (HMM) test configuration – current waveform probe verification. Permission granted from On Semiconductors, Inc.

Current Probe Waveform Comparison

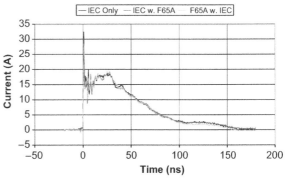

Figure 4.21   Human metal model (HMM) test configuration   current waveform probe comparison. Permission granted from On Semiconductors, Inc.

Figure 4.22 shows the test system configuration, where the source is not an ESD gun. The method applies an IEC pulse to the device under test without any air discharge. Using a current source, variations in the ESD gun waveform, and pulse variation is removed.

4.11 CHARGED BOARD MODEL (CBM)

With portable devices and growth of the cell phone market, the ESD sensitivity of semi-conductor chips mounted on small system boards is of growing interest. The charged board

Figure 4.22 Human metal model (HMM) test systems. Permission granted from Grund Technical Solutions LLC

model (CBM) represents the case of a semiconductor chip and a system board (also referred to as the charged board event (CBE)). In the CBM event, the semiconductor chip is mounted on the system board. The board and components are all charged through a V_{DD} or V_{SS} ground connection (similar to the un-socketed CDM test). The board and component are charged to a designated voltage with an external charging source. The board and mounted component are placed on top of an insulating surface, and a ground plate. The capacitance of the combined board and the component to the ground plane is larger than the component itself. As a result, the total charge stored in the board and component is larger than the charge that would have been stored with the component by itself. In the test procedure, any point on the board can be grounded; unlike the CDM test, where only the package signal pins are grounded. Additionally, unlike the human metal model (HMM) that only addresses external ports, in this test, any exposed physical point on the board can be grounded. Figure 4.23 shows the charged board model configuration.

4.12 CABLE DISCHARGE EVENT (CDE)

Cable discharge event (CDE) is an increasing concern in systems of all different physical scales. Charged cables are a concern with large scale computer systems, laptops, hand-held devices in the disk drive industry [18–25]. In large computer systems, CDE were controlled

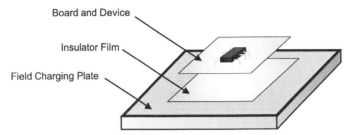

Figure 4.23   Charged board model configuration

by rigid procedures, wrist straps, and "touch pad" buttons to discharge the cables prior to insertion or electrical connections; these procedures are not possible in today's environment of laptops, cell phones, and mobile devices. System level engineers are required to improve system-level performance while maintaining the quality and reliability of the electronic system. ESD and EMI are a concern in systems since they can influence the visual interface (e.g. terminal, flat panel display), keyboards, system boards and electronics mounted on the system boards.

Electrical discharge from charged cables has been a concern in electrical systems. Electrical cable behaves as transmission lines having both capacitive and inductive characteristics. The charged stored in an electrical cable can be evaluated according to a stored capacitance per unit length. As the length of the cable increases, the amount of stored charge in the cable increases. In an "un-socketed" cable, the charge is stored between the center and outer conductors across the insulator region. When the voltage across the insulator exceeds the insulator breakdown voltage, an electrical discharge occurs between the outer and inner conductor. In the case when the voltage remains below the breakdown voltage, the stored charge remains in the cable, leaving the cable as a charged cable. When the charged cable center conductor approaches a system, an electrical arc occurs between the center conductor and the electrical system input signals. With the cable discharge event to the system level inputs, a significant current can discharge to the signal pin leading to CMOS latchup in the semiconductor chips integrated into the system components. The length of the applied current pulse is a function of the length of the cable.

As an example of how a common cable discharge phenomenon occurs, we can discuss the un-terminated twisted pair (UTP) cables. Charge accumulation on un terminated twisted pair cables occurs through both tribo-electric charging and induction charging. In the case of tribo-electrification, an un-terminated twisted pair cable is dragged along a floor surface. A positive charge is established on the outside surface of the insulating film. The positive charge on the outside of the cable attracts negative charge in the twisted pair leads across the dielectric region. When the negative charge is induced near the outside positive charge, positive charge is induced in the electrical conductor at the ends of the cable. As the cable is plugged into a connector, electrical arcing will occur leading to a charging of the un-terminated twisted pair (note: the twisted pair was neutral to this point).

A second charging process is induction charging. If a cable is introduced into a strong electric field, induction charging will occur. When the electric field is removed the cable remains charged until a discharge event from grounding occurs.

Historically, the CDE pulse in the system environment was resolved by operators discharging the charged cable prior to insertion into the system. This was achieved by handling procedures where system operators had "touch pads" to discharge the cable prior to insertion into the electronic system. Additionally, system level requirements may have been in place which did not allow "hot socketing" into large computer systems. In today's environments, systems are dynamically re-configurable, allowing power-up and power-down of sub-systems without turning off the system power. The requirement to "hot-plug" with the system powered as a system level requirement is quite common (e.g. also known as "fail-safe). Today's systems are portable units from small electronic systems: laptops to small servers. In these cases, the handling procedures of cables, cable connectors and interconnects are not rigorously followed.

4.12.1 Cable Discharge Event (CDE) and Scaling

As the number of electronic circuits increases, the number of I/O ports also increases (e.g. Rent's Rule). As the number of I/O ports increase, there is an increase in the number of electrical cable connections. As a result, future systems have an increasing number of electrical cable connections. This increases the likelihood of a charged cable connection leading to CMOS latchup. With the increase in the number of cables, and electrical mobility and reconfiguration, there is a higher incident of dis-connections and re-connections in the general usage of a system. Additional to the system level issues, the CMOS latchup robustness of advanced technology are significantly lower due to technology scaling of the latchup critical parameters. Given that the number of incidents has increased, and the CMOS latchup robustness of the technologies has decreased, there is a higher probability that CDE can lead to CMOS latchup of components. Hence, with both system and technology evolution, the reasons for the increased concern for this issue are as follows:

• Wide Area Networks (WAN), and Local Area Network (LAN) integration.

• Category 5 and 6 Local Area Network (LAN) Cabling.

• Higher level incidents of dis-connection.

• High level incidents of re-connections.

• Competitive focus on cost reduction and the elimination of future latchup solutions.

Figure 4.24 shows an example of the cable discharge event (CDE) pulse waveform.

4.12.2 Cable Discharge Event (CDE) – Cable Measurement Equipment

For accurate system measurements, it is important to have measurement equipment to capture the response. Figure 4.25 is the CDE measurement equipment used to capture the response reference waveform. Figure 4.26 is an example of the constant impedance transmission line adapter.

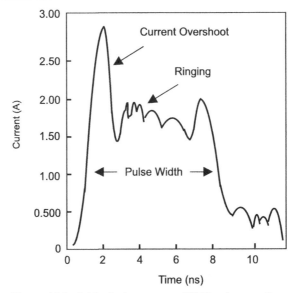

Figure 4.24 Cable discharge event (CDE) pulse waveform

To capture the cable waveform, a 50 Ω target transmission line adapter is utilized (Figure 4.27).

Cable discharge waveforms are a function of many variables. Cable discharge events are a function of the following:

• Capacitance per unit length.

• Cable length.

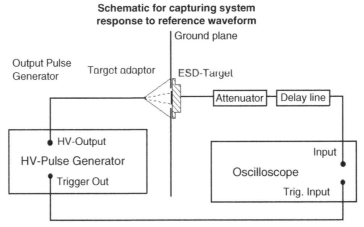

Figure 4.25 CDE measurement equipment – captured response for reference waveform. Permission granted from the ESD Association

**Constant Impedance Tapered Transmission
Line adapter**

50 Ohm conical adapter line   ESD current target

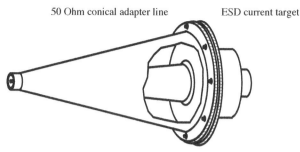

Figure 4.26 Constant impedance tapered transmission line adapter. Permission granted from Barth Electronics, Inc.

- Cable connector design.
- Cable configuration.

Cable configurations can be a cable on the ground, a cable taped to a wall, a cable held by a person, a "floating cable" and other configurations. In the following sections, examples will be discussed.

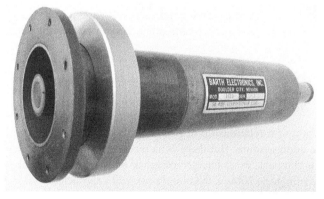

Figure 4.27   50 Ω Target transmission line adapter. Permission granted from Barth Electronics, Inc.

4.12.3 Cable Configuration – Test Configuration

In this study, the cable resides on the floor (a linoleum tile on concrete). A first vertical wall is placed perpendicular to the floor. A 100 meter cable is tested, where the first meter of the cable is held by and insulator, and the rest of the cable is stretched out on the floor to minimize coiling capacitance. The target is placed in an opening in the vertical metal ground plane. The oscilloscope is electrically connected to the target.

An example of a cable waveform response can be studied using an ESD gun source applied to the cable. The cable is measured with a Tektronix CT-3 current probe and a 20 dB attenuator. As anticipated, the waveform is an oscillatory decay response. With a Tektronix CT-1 current probe, the initial peak of the ESD gun pulse is captured.

4.12.4 Cable Configuration – Floating Cable

Modification of the cable position can influence the cable waveform. For example, when the first section of the cable near the vertical ground plane is free floating, the current is lower compared to a cable taped to the vertical ground plane. When the first meter of the cable is taped to the wall, there is a significant increase in the peak current.

4.12.5 Cable Configuration – Held Cable

A third configuration is when the first section of the cable is free floating and a person holds the cable near the target. The presence of the hand near the target also influences the cable waveform. The key point for testing cables and cable discharge events (CDE) is whether the cable configuration is coiled, uncoiled, free floating, taped, or held with a hand.

4.12.6 Cable Discharge Event (CDE) – Peak Current vs. Charged Voltage

The cable discharge event relationship of peak current and charge voltage is influenced by the different cable configuration. But there is a correlation between the charging voltage of the cable and the initial peak. For a charge voltage of 500 V, the initial peak is approximately 2.5 A. At a charge voltage of 2000 V, the initial peak will approach 10 A.

4.12.7 Cable Discharge Event (CDE) – Plateau Current vs Charged Voltage

Additionally, there is interest in the plateau section of the waveform. A linear relationship exists between the charge voltage and the plateau current. Hence quantification of the cable discharge event is possible for system level CDM testing.

4.13 SUMMARY AND CLOSING COMMENTS

In this chapter, system level concerns that are relevant to today's application and the future have been presented. A brief discussion of electrostatic issues in servers, laptops, hand-held devices, cell phones, disk drives, digital cameras, autos, and space applications was reviewed to educate the reader in the vast number of issues in today's electronic environment. System level ESD tests, such as IEC 61000-4-2, HMM, CDE and CBM were discussed.

In the next chapter, Chapter 5, electronic component solutions and design practices will be highlighted.

REFERENCES

1. Voldman, S. (2004) *ESD: Physics and Devices*, John Wiley and Sons, Ltd., Chichester, England.
2. Voldman, S. (2005) *ESD: Circuits and Devices*, John Wiley and Sons, Ltd., Chichester, England.
3. Voldman, S. (2006) *ESD: RF Circuits and Technology*, John Wiley and Sons, Ltd., Chichester, England.
4. Voldman, S. (2008) *ESD: Circuits and Devices*, Publishing House of Electronic Industry (PHEI), Beijing, China.
5. Voldman, S. (2009) *ESD: Failure Mechanisms and Models*, John Wiley and Sons, Ltd., Chichester, England.
6. Voldman, S. (2011) *ESD: Design and Synthesis*, Chichester, England, John Wiley and Sons, Ltd.
7. Voldman, S. (2007) *Latchup*, John Wiley and Sons, Ltd., Chichester, England.
8. Ker, M.D. and Hsu, S.F. (2009) *Transient Induced Latchup in CMOS Integrated Circuits*, John Wiley and Sons, Ltd., Singapore.
9. Mardiquan, M. (2009) Electrostatic Discharge, *Understand, Simulate, and Fix ESD Problems*, John Wiley and Sons, Co., New York.
10. Geski, H. (September 2004) DVI compliant ESD protection to IEC 61000-4-2 level 4 standard. *Conformity*, pp. 12–17.
11. International Electro-technical Commission (IEC) IEC 61000-4-2 (2001) Electromagnetic Compatibility (EMC): Testing and Measurement Techniques – Electrostatic Discharge Immunity Test.
12. Grund, E., Muhonen, K., and Peachey, N. (2008) Delivering IEC 61000-4-2 current pulses through transmission lines at 100 and 330 ohm system impedances. Proceedings of the Electrical Overstress/Electrostatic Discharge (EOS/ESD) Symposium, pp. 132–141.
13. IEC 61000-4-2 (2008) Electromagnetic Compatibility (EMC) – Part 4-2: Testing and Measurement Techniques – Electrostatic Discharge Immunity Test.
14. Chundru, R., Pommerenke, D., Wang, K. *et al.* (2004) Characterization of human metal ESD reference discharge event and correlation of generator parameters to failure levels – Part I: Reference Event. *IEEE Transactions on Electromagnetic Compatibility*, **46** (4), 498–504.
15. Wang, K., Pommerenke, D., Chundru, R. *et al.* (2004) Characterization of human metal ESD reference discharge event and correlation of generator parameters to failure levels – Part II: Correlation of generator parameters to failure levels. *IEEE Transactions on Electromagnetic Compatibility*, **46** (4), 505–511.
16. ESD Association ESD SP 5.6 2008 (2008) ESD Association Standard Practice for the Protection of Electrostatic Discharge Sensitive Items - Electrostatic Discharge Sensitivity Testing – Human Metal Model (HMM) Testing Component Level. Standard Practice (SP) document.
17. ANSI/ESD SP5.6-2009 (2009) Electrostatic Discharge Sensitivity Testing - Human Metal Model (HMM) - Component Level.

18. Intel Corporation (July 2001) Cable discharge event in local area network environment. White Paper, Order No: 249812-001.

19. Brooks, R. (March 2001) A simple model for the cable discharge event. *IEEE802.3 Cable-Discharge Ad-hoc Committee.*

20. Telecommunications Industry Association (TIA) (December 2002) Category 6 Cabling: Static discharge between LAN cabling and data terminal equipment, *Category 6 Consortium.*

21. Deatherage, J. and Jones, D. (2000) Multiple factors trigger discharge events in Ethernet LANs. *Electronic Design,* **48** (25), 111–116.

22. Stadler, W., Brodbeck, T., Gartner, R., and Gossner, H. (2006) Cable discharges into communication interfaces. Proceedings of the Electrical Overstress/Electrostatic Discharge (EOS/ESD) Symposium, pp. 144–151.

23. ESD Association DSP 14.1-2003 (2003) ESD Association Standard Practice for the Protection of Electrostatic Discharge Sensitive Items – System Level Electrostatic Discharge Simulator Verification Standard Practice. Standard Practice (SP) document.

24. ESD Association DSP 14.3-2006 (2006) ESD Association Standard Practice for the Protection of Electrostatic Discharge Sensitive Items – System Level Cable Discharge Measurements Standard Practice. Standard Practice (SP) document.

25. ESD Association DSP 14.4-2007 (2007) ESD Association Standard Practice for the Protection of Electrostatic Discharge Sensitive Items – System Level Cable Discharge Test Standard Practice. Standard Practice (SP) document.

26. Jowett, C.E. (1976) *Electrostatics in the Electronic Environment,* Halsted Press, New York.

27. Lewis, W.H. (1995) *Handbook on Electromagnetic Compatibility,* Academic Press, New York.

28. Morrison, R. and Lewis, W.H. (1990) *Grounding and Shielding in Facilities,* John Wiley and Sons Inc., New York.

29. Paul, C.R. (2006) *Introduction to Electromagnetic Compatibility,* John Wiley and Sons Inc., New York.

30. Morrison, R. and Lewis, W.H. (2007) *Grounding and Shielding,* John Wiley and Sons Inc., New York.

31. Ott, H.W. (2009) *Electromagnetic Compatibility Engineering,* John Wiley and Sons Inc., Hoboken, New Jersey.

32. Ott, H.W. (1985) Controlling EMI by proper printed wiring board layout. Sixth Symposium on EMC, Zurich, Switzerland.

33. ANSI C63.4-1992 (July 17 1992) *Methods of Measurement of Radio-Noise Emissions from Low-Voltage Electrical and Electronic Equipment in the Range of 9 kHz to 40 GHz,* IEEE.

34. EN 61000-3-2 (2006) *Electromagnetic Compatibility (EMC) – Part 3-2: Limits-Limits for Harmonic Current Emissions (Equipment Input Current < 16 A Per Phase),* CENELEC.

35. EN 61000-3-3 (2006) *Electromagnetic Compatibility (EMC) – Part 3-3: Limits-Limitation of Voltage Changes, Voltage Fluctuations and Flicker in Public Low-Voltage Supply Systems for Equipment with Rated Current < 16A Per Phase and Not Subject to Conditional Connection,* CENELEC.

36. EN 61000-4-2 (2001) Electromagnetic Compatibility (EMC) – Part 4-2: Testing and Measurement Techniques – Electrostatic Discharge Immunity Test.

37. MDS MDS-201-0004 (October 1 1979) *Electromagnetic Compatibility Standards for Medical Devices,* U.S. Department of Health Education and Welfare, Food and Drug Administration.

38. MIL-STD-461E (August 20 1999) *Requirements for the Control of Electromagnetic Interference Characteristics of Subsystems and Equipment.*

39. Radio Technical Commission for Aeronautics (RTCA) RTCA/DO-160E (December 7 2004) *Environmental Conditions and Test Procedures for Airborne Equipment,* Radio Technical Commission for Aeronautics (RTCA).

40. Society of Automotive Engineers SAE J551 (June 1996) Performance Levels and Methods of Measurement of Electromagnetic Compatibility of Vehicles and Devices (60 Hz to 18 GHz).

41. Society of Automotive Engineers SAE J1113 (June 1995) *Electromagnetic Compatibility Measurement Procedure for Vehicle Component (Except Aircraft) (60 Hz to 18 GHz)*, Society of Automotive Engineers.

42. Wall, A. (2004) Historical perspective of the FCC rules for digital devices and a look to the future. IEEE International Symposium on Electromagnetic Compatibility.

43. Denny, H.W. (1983) *Grounding For the Control of EMI*, Don White Consultants, Gainesville, VA.

44. Boxleitner, W. (1989) *Electrostatic Discharge and Electronic Equipment*, IEEE Press, New York.

45. Gerke, D.D. and Kimmel, W.D. (March/April 1986) Designing noise tolerance into microprocessor systems. *EMC Technology*.

46. Kimmel, W.D. and Gerke, D.D. (September 1993) Three keys to ESD system design. *EMC Test and Design*.

47. Violette, J.L.N. (May/June 1986) ESD case history – Immunizing a desktop business machine. *EMC Technology*.

48. Wong, S.W. (October 1984) ESD design maturity test for a desktop digital system. *Evaluation Engineering*.

5 Component Level Issues – Problems and Solutions

A semiconductor chip was assembled with a unique package design, which had large inductance from the package pin to the location of the circuitry within the "spine" of the chip. Bond pads, off-chip drivers, and receivers were not placed on the edges but along the center axis of the chip.

One design was functional on wafer, but not after packaging the chip due to the lead frame inductance. It was decided to separate the power rails of the chip between core and peripheral circuitry due to the functionality problem. In this transition, six ESD failure mechanisms were introduced. So, by fixing the functionality, the chip architecture was modified, leading to ESD failures. A new ESD device was needed to address this problem.

5.1 ESD CHIP PROTECTION – THE PROBLEM AND THE CURE

Component level ESD protection is achieved using "on-chip" ESD protection circuits in integrated circuits [1–10]. ESD protection circuits are integrated into the semiconductor chip design, providing an alternative current path for discharging the ESD current. Circuitry along the signal path of a semiconductor chip are sensitive to high current and high voltage phenomena.

ESD protection networks provide an alternative current path to shunt the ESD current to the power supply rail, or to the ground power rails [2–7]. ESD protection circuitry is placed on signal pins, and power pins to establish these current paths. ESD signal pin protection circuitry introduces a means to allow the current to be diverted from the signal path to the power rails. ESD protection circuits are also placed between the power rails (e.g., the power

ESD Basics: From Semiconductor Manufacturing to Product Use, First Edition. Steven H. Voldman.
© 2012 John Wiley & Sons, Ltd. Published 2012 by John Wiley & Sons, Ltd.

supply rail V_{DD}, and ground rail V_{SS}) to complete the Kirchoff current loop to the grounded referenced rail. These networks are known as "ESD power clamp" circuitry.

In today's semiconductor components, ground rails are separated to avoid noise interference with the circuitry. Circuits are separated into different power domains to isolate the influence of one circuit group impacting a second [7]. Different power domains separate both the power supply rails (e.g., V_{DD}) and ground rails (e.g., V_{SS}). A third class of ESD circuitry is placed between the power supply rails, and between the ground rails of the different domains to allow current paths during ESD events [7].

5.2 ESD CHIP LEVEL DESIGN SOLUTIONS – BASICS OF DESIGN SYNTHESIS

ESD chip level design solutions can be separated into different classes of the circuitry. In the following section we will discuss the basics of ESD design synthesis. The section will first discuss the ESD circuits, followed by discussion of chip architecture, power rails, and other design considerations.

As a design discipline, the ESD design discipline is distinct from circuit design practices used in the development of semiconductor circuit design discipline [2,3]. Fundamental concepts and objectives exist in the electrostatic discharge (ESD) design of semiconductor devices, circuits and systems in methods, layout, to design synthesis.

To address the first issue of the ESD design discipline, let us first address the distinction of ESD design discipline practice uniqueness. Here are some of the ESD design practices [2,3]:

- Device Response to External Events: Design of devices and circuits to respond to (and not to respond to) unique current waveforms (e.g., current magnitude and time constants) associated with external environments. In ESD design, the ESD devices as well as the circuits which are to be protected can be designed to respond to (and not to respond to) unique ESD current waveforms. ESD networks typically are designed to respond to specific ESD pulses. These networks are unique in that they address the current magnitude, frequency, polarity and location of the ESD events. Hence, in ESD design, the ESD networks are designed and tuned to respond to the various ESD events. In ESD design, different stages or segments of the network can also be designed to respond to different ESD events. For example, some stages of a network can respond to human body model (HBM) and machine model (MM) events, while other segments respond to the charged device model (CDM) event. These ESD events differ in current magnitude, polarity, time constant, as well as the location of the current source. Hence, the ESD circuit is optimized to respond to and address different aspects of ESD events that circuits may be subjected to [2,3].

- Alternate Current Loops: Establishment of alternative current loops or current paths which activate during high current or voltage events. A unique issue is the establishment of alternative current loops or current paths which activate during high current or voltage events. By establishing alternative current loops, or secondary paths, the ESD current can be re-directed to prevent over-voltage of sensitive circuits. In order to have an effective

ESD design strategy, this current loop must respond to the ESD event and have a low impedance [2,3].

- Switches: Establishment of "switches" that initiate during high current or voltage events. On the issue of establishment of "switches" that initiate during high current or voltage events, the uniqueness factor is that these are either passive or activated by the ESD event itself. A unique feature of ESD design is that it must be active during un-powered states. Hence, the "switches" used to sway the current into the ESD current loop are initiated passively, or are initiated by the ESD event itself. Hence, the ESD event serves as the current and voltage source to initiate the circuit. These switches lead to "current robbing" and the transfer of the majority of the current from the sensitive circuit to the alternative current loop. The ESD design discipline must use "switches" or "triggers" that initiate passively (e.g., a diode element), or actively (e.g., a frequency-triggered ESD network). A design objective is to provide the lowest voltage trigger allowable in the application space. Hence, a key ESD design objective is to utilize low voltage trigger elements that serve as a means to transfer the current away from the sensitive circuit to alternative current paths. A large part of effective ESD design discipline is the construction of these switches or trigger elements [2,3].

- Decoupling of Current Paths: De-coupling of sensitive current paths is an ESD design discipline practice. Circuit elements can be introduced which lead to the avoidance of current flow to those physical elements. The addition of "ESD de-coupling switches" can be used to de-couple sensitive circuits as well as avoid the current flow to these networks or sections of a semiconductor chip. ESD de-coupling elements can be used to allow elements to undergo open or floating states during ESD events. This can be achieved within the ESD network, or within the architecture of a semiconductor chip. Decoupling of sensitive elements or decoupling of current loops can be initiated by the addition of elements that allow the current loop to "open" during ESD events. The de-coupling of nodes, elements, circuits, chip sub-functions, or current loops relative to the grounded reference prevents over-voltage states in devices, and eliminates undesired current paths. De-coupling elements can avoid "pinning" of electrical nodes. Hence, integration of devices, circuit elements or circuit function that introduce de-coupling electrical connections to ground references, and power supplies references, is a key ESD design practice [2,3].

- Decoupling of Feedback Loops: De-coupling of loops that initiate pinning during off condition or ESD test modes. Feedback loops can lead to unique ESD failures and lower ESD results significantly. The de-coupling of nodes, elements or current loops relative to the grounded reference prevents over-voltage states in devices, and eliminates current paths initiated by the feedback elements. These de-coupling elements can avoid "pinning" of electrical nodes [2,3].

- Decoupling of Power Rails: De-coupling of electrical connections to grounded references, and power supplies [2,3].

- Local and Global Distribution: Local and global distribution of electrical and thermal phenomena in devices, circuits, and systems is a key ESD design practice and focus in ESD development. To provide an effective ESD design strategy, the ESD design practices must focus on the local and global distribution of electrical and thermal phenomena in devices, circuits, and systems. In order to shunt the ESD current efficiently and effectively, the distribution of the current is critical in ESD design. As the current distributes, the effectiveness of the device improves the utilization of the total area of the ESD network or circuit element. On a circuit and system level, the distribution of the ESD current within the network or system, lowers the effective impedance, and lowers the voltage condition within the ESD current loop [2,3].

• Usage of Parasitic Elements: Utilization and avoidance of parasitic elements is part of the ESD design practice. ESD design either utilizes or avoids activation of these parasitic elements in the ESD implementations. Utilization of parasitic elements is a common ESD design practice for ESD operation, such as utilization of parasitic lateral or vertical bipolar transistors. It is not common to use these parasitic elements in standard circuit design, whereas for ESD design it is very prevalent to utilize the parasitic devices and is part of the ESD design practice and art [2,3].

• Buffering: Utilization of current and voltage buffering of sensitive devices, circuits or sub-circuits is a key ESD design practice. In ESD design, it is common practice to establish current and voltage buffering of sensitive devices, circuits, sub-circuits, chip level core regions, or voltage islands. A design practice is to increase the impedance in the path of the sensitive circuit either by placement of high impedance elements, establishing "off" states of elements, voltage and current dividing networks, resistor ballasting or initiating elements in high impedance states [2,3].

• Ballasting: It is common ESD design practice to use ballasting techniques. Introduction of resistance to re-distribute current within a single element or a plurality of elements. In digital design, ballasting is predominately achieved using resistor elements. Resistive, capacitive or inductive ballasting can be introduced to re-distribute current or voltage within a single element or a plurality of elements, circuit, or chip segment. The usage within a semiconductor device element allows for re-distribution within a device to avoid electro-thermal current constriction, and poor area utilization of a protection network or circuit element. The use of ballasting allows the source current from the ESD event to be redistributed to avoid thermal heating or electrical overstress within the semiconductor network or chip. Ballasting can be introduced into a semiconductor device structure achieved by semiconductor process choices, material choices, silicide film removal, intro duction of discrete resistor elements, and introduction of design layout segmentation [2,3].

• Usage of Unused Sections of a Semiconductor Device, Circuit or Chip Function: It is an ESD design practice to utilize "unused" segments of a semiconductor device for ESD protection, which was not utilized for functional applications [2,3].

• Impedance Matching between Floating and Non-Floating Networks: It is ESD design practice to impedance match the states of floating structures. In ESD design, it is common to utilize the "unused" segments of a semiconductor device for ESD protection and impedance match the network segments for ESD operation; this matching of conditions during ESD testing allows for current sharing during matching between networks, and common triggering voltage conditions [2,3].

- Unconnected Structures: It is common ESD design practice to address structures not containing electrical connections to the power grid or circuitry. In semiconductor chips, there are many structures which are electrically not connected to other circuitry or power grids which are vulnerable to ESD damage. Unique ESD solutions are used to address floating or unconnected structures [2,3].

- Utilization of "Dummy Structures and Dummy Circuits": In ESD design practice it is not uncommon to utilize dummy structures or dummy circuits which serve the purpose of providing better current uniformity or distribution effects; this concept spans from using dummy MOSFET polysilicon gate fingers to dummy inverter circuits [2,3].

- Non-scalable Source Events: Another key issue is that the ESD event is a non-scalable event. With each generation, the size of devices are scaled to smaller dimensions. The ESD design practice must address the constant source input current, and the physical scaling of the structures. A unique ESD scaling theory and strategy must be initiated to address this issue [2,3].

5.2.1 ESD Circuits

ESD circuits can exist externally and internally to a semiconductor chip. The most common use is of ESD networks associated with the bond pads due to the external sources. Today, with separated power domains, there are ESD failures associated with power domains in system-on-chip (SOC) applications.

5.2.2 ESD Signal Pin Protection Networks

For a signal pin ESD network, a pnpn silicon controlled rectifier was introduced into a technology on a p-/p++ wafer. Due to epitaxial control issues, the SCR ESD circuit trigger voltage and response was a function of the incoming epitaxial thickness, and epitaxial resistance. This chip was fabricated in three different locations where the wafers and epitaxial thickness measurement methodology was not equivalent. The ESD results succeeded in one plant, but failed at the other two locations.

The purpose of the ESD signal pin protection networks is to divert the current from the signal path to the power rails. The ESD signal pin protection networks must be placed prior to the signal pin functional circuitry. As a result, the ESD signal pin protection networks are typically near the signal pin bond pad, and in many cases integrated into the peripheral circuitry layout design and circuitry.

ESD signal pin protection networks are bi-directional to allow current to flow from both positive or negative polarity events. In many cases, the circuitry bi-directionality is not apparent, but addressed through a parasitic element.

In Figure 5.1, a typical ESD network used for ESD protection is shown. This network is known as the dual diode ESD protection network. In this ESD protection circuit, a first element is electrically connected to the V_{DD} power supply rail, and the second element is

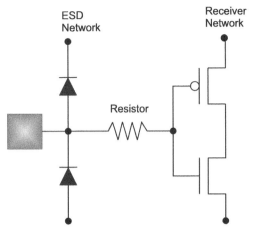

Figure 5.1   ESD signal pin network – the dual diode ESD network

electrically connected to the V_{SS} ground rail. For this ESD signal pin protection network, there exists a forward bias diode element for a positive or negative polarity pulse to divert the current to the V_{DD} and V_{SS} power rails.

In Figure 5.2, a second typical ESD network used for ESD protection is shown. This network is known as the GGNMOS ESD protection network. GGNMOS is an acronym for a grounded gate n-type MOSFET device ESD network. In this ESD protection circuit, only one element exists, which is electrically connected to the V_{SS} ground rail. For this ESD signal pin protection network, there exists a "snapback" event for positive polarity events allowing discharge to the V_{SS} power rail, and forward bias diode for negative polarity events. In the example shown, a circuit is shown that contains a two-stage high voltage and a low voltage GGNMOS solution, which was used primarily in the 1980s. As technology advanced, typically only a single-stage low voltage GGNMOS was used.

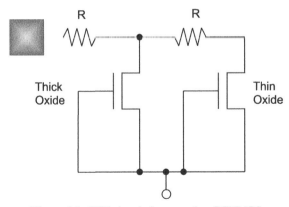

Figure 5.2   ESD signal pin network – GGNMOS

5.2.3 **ESD Power Clamp Protection Networks**

ESD power clamp networks are ESD circuits which are placed between the power supply V_{DD}, and ground power rail V_{SS} to establish an alternate current loop for the ESD current. ESD power clamp networks consist of two fundamental features. ESD power clamps typically contain a "trigger" network, and a "clamp" or "shunt" network. The "trigger network" keeps the ESD power clamp "off" during power up, functional usage and power down. During an overvoltage state, or ESD event, the ESD power clamp is "on" to discharge the ESD current.

Figure 5.3 is an example of an ESD power clamp used throughout the semiconductor industry. This circuit is known as an RC-triggered ESD MOSFET power clamp. The trigger network is an RC discriminator network that can distinguish between an ESD event, and non-ESD event. This circuit is normally "off" during semiconductor chip operation. The RC circuit is tuned to respond to the human body model (HBM) pulse event. When an HBM event occurs, the RC network initiates the inverter drive stage to turn on the ESD power clamp MOSFET element [2–10].

Figure 5.4 is an example of an ESD power clamp used in bipolar and BiCMOS technology. This circuit is known as a BVCEO-triggered ESD bipolar power clamp. The trigger network is a transistor which turns on when the collector-to-emitter breakdown voltage is established across the trigger element. This turns on the transistor, which then provides base current to the ESD clamp element. A distinction between the first and second example is that the first is "frequency triggered" and the second is "voltage triggered." Depending on the application, there are advantages and disadvantages of these two means of initiating the ESD power clamp circuit [2–5].

5.2.4 **ESD Power Domain-to-Domain Circuitry**

ESD circuits are also placed between ground rails of separate power domains. In the early 1980s and 1990s, it was common to use ESD networks between V_{DD} power supplies. Today, it is common practice to place these networks between the ground power rails. Figure 5.5 is

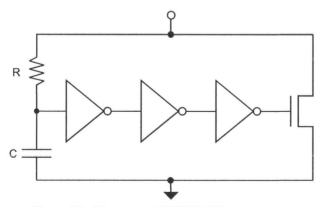

Figure 5.3 RC-triggered MOSFET ESD power clamp

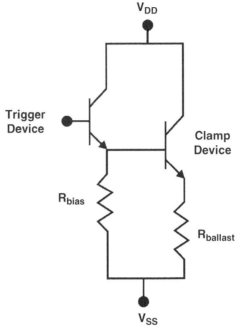

Figure 5.4   BVCEO-triggered ESD bipolar power clamp

an example of an ESD network placed between an analog VSS and digital VSS. This network consists of p-n diode elements. Note that the network is bi-directional [2–8].

5.2.5 ESD Internal Signal Line Domain-to-Domain Protection Circuitry

ESD networks typically were only required on signal pins and power rails, and not on internal signal lines. In mixed signal applications, digital and analog power domains are separated to prevent the digital noise from influencing the analog circuitry. Signal lines from the digital on-chip driver circuitry to the analog receiver networks can be the only connectivity between the two domains (Figure 5.6). In recent years, ESD failures have occurred on these signal lines due to MOSFET gate over-voltage [6,7].

A solution to address the MOSFET over-voltage on the receiver was to introduce "internal ESD networks" on the analog receiver. Figure 5.7 is an example of an internal signal line

Figure 5.5   ESD domain-to-domain power rail network

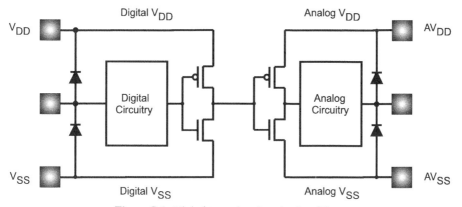

Figure 5.6   Digital to analog domain signal lines

ESD network. On the internal signal line between the digital and analog domain, a resistor and ground gate MOSFET (e.g., GGNMOS) circuit elements are placed to prevent overvoltage of the analog MOSFET receiver [7].

5.3 ESD CHIP FLOOR PLANNING – BASICS OF DESIGN LAYOUT AND SYNTHESIS

In the design of semiconductor chips, the physical placement of the ESD networks within the semiconductor design can influence the ESD robustness of the semiconductor chip. In this section, the placement of the elements in the semiconductor chip will be discussed.

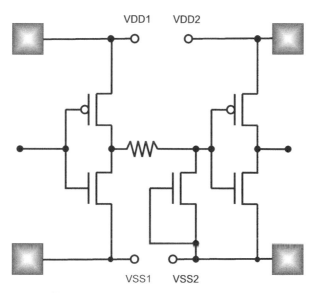

Figure 5.7   Internal signal line ESD network

5.3.1 Placement of ESD Signal Pin HBM Circuitry

In the design of semiconductor chips, the physical placement of the ESD signal pin networks are typically adjacent to the bond pad of the signal pin [7]. It is a misconception that this is a necessary requirement for good ESD protection. In most cases, it is a convenient location, but not mandatory requirement. In peripheral I/O design, typically the ESD signal pin network is placed in the periphery of the chip near the bond wire, and bond pad. Figure 5.8 shows an example of an ESD network placed within a peripheral I/O standard cell.

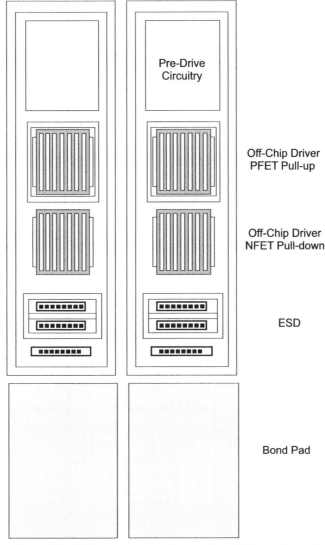

Figure 5.8 ESD network placed in a peripheral I/O standard cell

In high pin-count semiconductors, this may not be true. The relative location of the bond pad and the ESD network may be spatially separated, and interconnected with a "transfer wire" to connect from the solder ball/bond pad structure and the ESD and I/O circuit [7].

5.3.2 Placement of ESD Signal Pin CDM Circuitry

In signal pin receiver networks, charged device model (CDM) events can lead to MOSFET gate oxide failure. The relative spatial location of the receiver and the ESD network will influence the ESD protection from CDM events. During the CDM charging process, charge is distributed through the entire chip substrate. If the receiver network is spatially separated from the ESD network, current may flow through the MOSFET gate dielectric instead of through the semiconductor chip substate. When a thin oxide receiver is far from the ESD network, then it is necessary to add a secondary stage ESD network. This secondary stage network is referred to as a CDM protection circuit. In the placement of the CDM circuit, the CDM circuit must be local or adjacent to the receiver network. Figure 5.9 shows an example of a CDM protection network commonly used.

5.3.3 Placement of ESD Power Clamp Circuitry

A key metric of ESD design is the power bus resistance between the signal pin and the ESD power clamp. In semiconductor chip design, where the power bus resistance is low, and the semiconductor chip size is small, it is possible to place the ESD power clamps in a convenient location that does not impact chip size. Historically, in many designs, the corners of the semiconductor chips were not being utilized for active circuits or peripheral I/O networks. As a result, this was a convenient location to place the ESD power clamps. Figure 5.10 shows an example of a semiconductor floor plan with the ESD power clamps placed in the corners of the design [7].

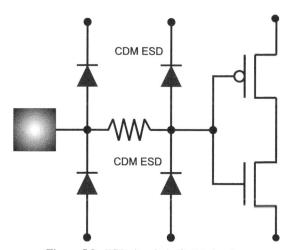

Figure 5.9 ESD signal pin CDM circuitry

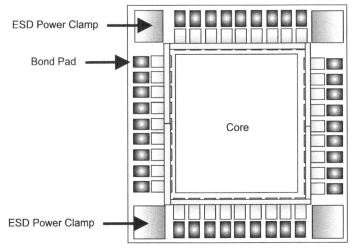

Figure 5.10   Floor plan for ESD power clamps – corner implementation

As semiconductor chip sizes increase, and the width of the power bus are decreased, the placement of ESD power clamps was inadequate due to a high resistance between any signal pin and its closest ESD power clamp. In many architectures, V_{DD} and V_{SS} power pins are placed at a high periodicity near signal pins. As a result, it became a convenient location to place the ESD V_{DD}-to-V_{SS} power clamps in the power pads (e.g., V_{DD}, or V_{SS}) of the semiconductor chip (Figure 5.11) [7].

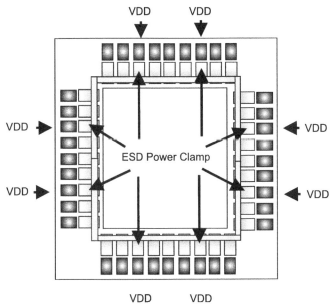

Figure 5.11   ESD power clamp floorplan – distributed implementation in V_{DD} power pads

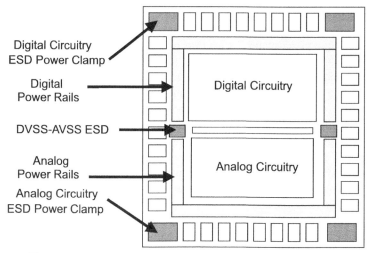

Figure 5.12 Floorplan for placement of DV_{SS}-AV_{SS} ESD networks

5.3.4 Placement of ESD V_{SS}-to-V_{SS} Circuitry

In semiconductor chip floor planning, the different circuit domains are separated due to different voltage requirements as well as to provide noise isolation. For example, analog and digital power domains are separated spatially on the semiconductor chip. Figure 5.12 shows an example where the digital and analog circuitry is spatially and electrically separated from each other. In this type of semiconductor chip architecture, the power rails are also separated. From the figure, where the power rails of the different domains are truncated, it is a good location for analog ground-to-digital ground ESD networks. This network is a bi-directional network to connect analog ground (AV_{SS}) to digital ground (DV_{SS}) [7].

5.4 ESD ANALOG CIRCUIT DESIGN

In analog design, unique design practices are used to improve the functional characteristics of analog circuitry [11,12]. In the ESD design synthesis of analog circuitry, the ESD design practices must be suitable and consistent with the needs and requirements of analog circuitry [7,11,12]. Fortunately, many of the analog design practices are aligned with ESD design practices.

In the analog design discipline, there are many design techniques to improve tolerance of analog circuits. Analog design techniques include the following [11]:

- Local Matching. Placement of elements close together for improved tolerance.
- Global Matching: Placement in the semiconductor die.
- Thermal Symmetry: Design Symmetry.

A key analog circuit design requirement is matching. To avoid semiconductor process variations, matching is optimized by the local placement. Placement within the die location is also an analog concern due to mechanical stress effects. In analog design, there is a concern of the temperature field within the die, and the effect of temperature distribution within the die.

Many of the analog design synthesis and practices are also good ESD design practices. The design practice of matching and design symmetry is also suitable for electrostatic discharge design. But, there are some design practices where a tradeoff exists between the analog tolerance and ESD; this occurs when parasitic devices are formed between the different analog elements within a given circuit, or circuit-to-circuit.

In the analog ESD discipline, there are a wide number of ESD devices and unique applications. There are CMOS, BiCMOS, BCD, SOI and smart power applications with a plethora of power supply voltage conditions and unique applications. In this section, we will just discuss one case and go in depth, and provide the reader some insight to just one small issue for analog ESD design.

A challenge for today's analog and radio frequency (RF) applications is to provide differential pair receiver circuitry. The challenges today in differential pair circuitry are:

- Matching.

- Low Capacitance.

For ESD protection, the challenges are twofold:

- Signal pin-to-rail ESD protection.

- Differential Pair Pin-to-Pin ESD protection.

As semiconductor products are scaled, power bus and electrical connections to ESD networks are scaled in ASICs, standard cell foundry design and memory products. In the past, bus resistance was an ESD design issue between the ESD input circuit element and the ESD power clamp element.

5.4.1 Symmetry and Common Centroid Design for ESD Analog Circuits

One solution to improve these issues is the use of common centroid design practices as well as parasitic elements. The concept of using a common centroid for the circuit, the signal pin-to-rail ESD, and then co-design and co-synthesis of the differential pair pin-to-pin ESD protection is discussed. Symmetry is important for minimizing design variation for both circuits and ESD networks. A metric to define symmetry is by establishing an axis of symmetry. Figure 5.13 demonstrates a common centroid where the circuit can be an analog differential pair network.

Common centroid design introduces four rules: (1) Coincidence, (2) Symmetry, (3) Dispersion, and (4) Compactness. To minimize variations, symmetry can be evaluated in one dimension, or both dimensions. For example, an axis of symmetry can be defined in the

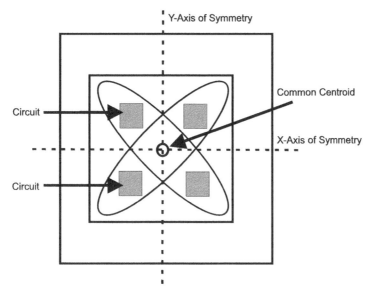

Figure 5.13 Common centroid design

x-axis and a second axis of symmetry can be defined in the y-axis. From this axis of symmetry, a common centroid can be established. Common centroid design is a process used in analog design practices.

5.4.2 Analog Signal Pin to Power Rail ESD Network

ESD signal pin-to-rail protection networks impact the mismatch and loading capacitance of the differential pair circuitry.

A first critical issue is the mismatch introduced between the two sides of the differential pair from the ESD structure itself. With the addition of the ESD network on the IN (+) and second ESD network on IN (−), a mismatch occurs since both ESD networks are spatially separated. Hence, the spatial separation of the two separate ESD networks can lead to functional implications in itself.

A second issue, given that either diode, MOSFET or SCR ESD networks are used, is that the loading capacitance of the ESD network impacts differential pair receiver performance. In the case of a dual diode ESD network, two additional diode capacitances are added to the differential pair network for both IN(+) and IN(−). This common centroid concept is typically not extended to ESD networks.

5.4.3 Common Centroid Analog Signal Pin to Power Rail ESD Network

To provide improved matching, the two ESD networks of the differential pair can have a common centroid. With the ESD network on the IN (+) and second ESD network on IN (−), mismatch can be reduced with the ESD networks having a common centroid. A second novel step is to have the ESD networks for IN(+) and IN(−) share a common region and area. For example, the two ESD networks can share a common p-well or n-well region. For example, to minimize mismatch of p+/n-well junctions, the two ESD networks can share a common n-well shape. Standard practice is to place the ESD at the bond pads, which spatially separate the ESD networks.

Figure 5.14 is an example where the two differential ESD elements are placed local to each other and in a common n-well tub. Normally, ESD element for IN(+) is placed at the signal pad for IN(+) and ESD element for IN(−) is placed at the signal pad for IN(−). In this example, the ESD elements are placed in a common shared region. In this case, the n-well region is common, and the spatial separation between IN(+) and IN(−) is minimum. Figure 5.14 shows one of the two diodes (e.g., P+/NW diodes), where the two ESD networks are placed in a common tub and placed locally together in a common array. Note, the second N+/PW can be designed equivalently.

Figure 5.15 shows an example where the fingers of the ESD-to-rail network are in a common well, but also are interlaced. In an array, the finger can be alternating, *ad infinitum*.

5.4.4 Co-synthesis of Common Centroid Analog Circuit and ESD Networks

To provide improved matching, the two ESD networks of the differential pair can be co-synthesized with the differential circuit, and share a common centroid in one or two dimensions. With layout co-synthesis, the input devices and the ESD networks can share a common centroid in both the x-axis of symmetry and the y-axis of symmetry. For example, an NFET ESD network can be placed with the circuit itself (e.g., an NFET) in a common centroid fashion.

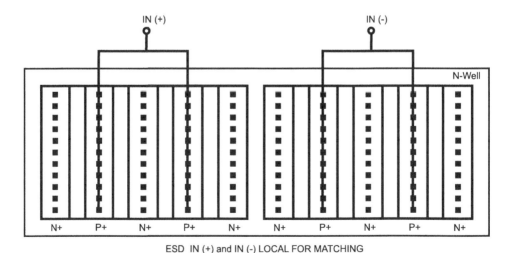

ESD IN (+) and IN (-) LOCAL FOR MATCHING

Figure 5.14 ESD pin-to-rail network pair with shared common well region layout

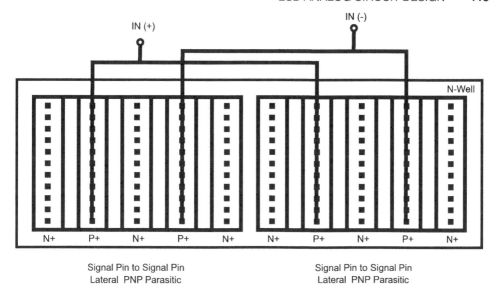

Figure 5.15 Cross section of ESD pin-to-rail network pair with shared common well region

5.4.5 Signal Pin-to-Signal Pin Differential Pair ESD Network

ESD signal pin-to-signal pin protection networks are required to provide ESD protection from the two pins within a differential pair circuit. Typically in both analog and RF applications (e.g., note: CMOS and bipolar), the differential pair pins are the most sensitive pins in a given semiconductor chip. It has also been shown that in "signal pin to all other signal pins (reference ground)" that the failure mechanism occurs between the two pins of the differential pair. Figure 5.16 shows a standard practice of a differential pair circuit with both ESD pin-to-rail protection network, and ESD pin-to-pin protection networks. With the addition of the differential pair pin-to-pin ESD networks, there is both an area and capacitance loading impact to the differential pair circuit performance. In addition, without a common centroid implementation, it will introduce a mismatch.

5.4.6 Common Centroid Signal Pin Differential Pair ESD Protection

For CMOS differential pair, there are two solutions for establishing the signal pin-to-signal pin ESD networks. Presently, the established conventional method is to utilize a bi-directional ESD network between the IN($+$) and IN($-$). Typically it is not a common centroid implementation.

A novel method is to introduce a common centroid implementation, where one ESD array serves both sides of the differential pair network. In this fashion, a common centroid design

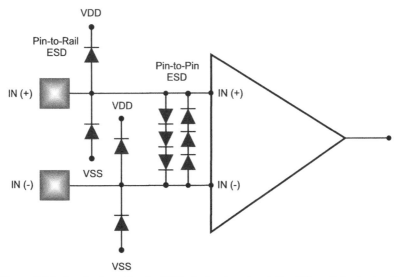

Figure 5.16   Standard practice for ESD pin-to-pin network for differential pair circuits

can be achieved. By taking the next step of alternating the fingers of the one ESD array, the parasitic elements between the two sides can be used (Figures 5.17 and 5.18).

For example, a lateral parasitic pnp can be formed between adjacent fingers of the differential pair ESD network p+/n-well diode. Additionally, a lateral parasitic npn can be used between adjacent fingers of the differential pair ESD network n+/p-well.

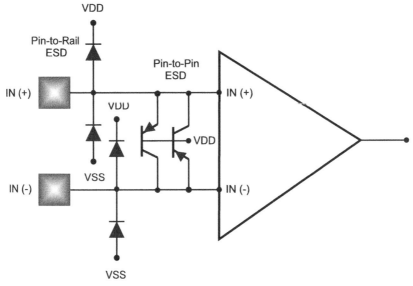

Figure 5.17   ESD pin-to-pin network for differential pair circuits utilizing pnp parasitics

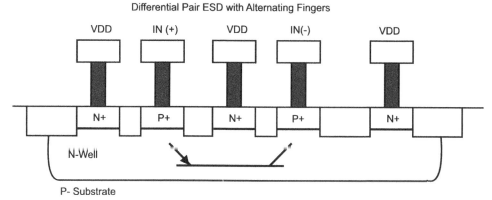

Figure 5.18 Cross section of ESD pin-to-pin network for differential pair circuits utilizing pnp parasitics. Note that a symmetric ESD-to-rail network is also utilized

This achieves multiple objectives:

- Common centroid design with improved matching.

- No additional capacitance load on the differential pair receiver network.

- No additional area for an additional ESD network.

In conclusion, the issue of common centroid design of ESD protection networks which integrates signal pin-to-signal pin ESD protection with the inter-digitation pattern for ESD pin-to-rail protection network for differential pair circuitry is discussed for the first time. With integration of the ESD pin-to-rail solution and the ESD signal pin solution, a significant reduction of the area and loading effect for CMOS differential circuits is established. This novel concept will provide significant advantage for present and future high performance analog and RF design for matching, area reduction, and performance advantages.

5.5 ESD RADIO FREQUENCY (RF) DESIGN

In the ESD design discipline of radio frequency (RF) circuits, there is a fundamental difference in the focus and methods that are required which is distinct from the ESD design practices used in ESD protection of digital circuitry [2,4,5,13]. This rapidly developing ESD design practice is utilizing some of the ESD digital design practice when it is possible and abandoning some practices, ESD circuits and design when they are unsuitable for RF applications. In this evolution, the ESD design discipline is shifting and adapting design practices used by microwave RF circuit designers.

5.5.1 ESD Radio Frequency (RF) Design Practice

A new unique RF ESD design practice has been established to co-synthesize RF functional application needs and ESD protection. To address the question of an RF ESD design discipline, we pose the following questions:

- What is it that makes the ESD design discipline unique?

- How is it distinct from standard circuit design practices?

- How is RF ESD design discipline different from the RF design discipline?

- How is RF ESD design discipline different?

The RF fundamental concepts for the ESD protection and design of radio frequency (RF) components are as follows [4,5]:

- RF ESD Application Frequency Dependent ESD Solutions: In RF ESD design, the solutions and methods for the ESD protection may be a function of the application frequency. Below 1 GHz, traditional digital ESD on-chip silicon ESD circuit solutions may be sufficient. Between 1 and 5 GHz, the choice of ESD device may be a function of the tradeoffs of loading and other RF parameters. Above 5 to 15 GHz, RF ESD co-synthesize may be a mandatory process. Above these application frequencies, off-chip protection and non-traditional ESD solutions may be necessary.

- RF Models for ESD elements: With RF circuits and components, d.c. and RF models are required to build RF circuits. As a result, all ESD elements must have full RF quality models. This is very different from ESD digital design practices that are not highly dependent on an ESD model. ESD design for digital design does not require a physical model. On the other hand, RF applications require some form of RF model analysis of the ESD element since it influences all the RF functional parameters. This influences the physical design implementation.

- RF ESD Design Methodology: With the requirement of high quality RF models, ESD design methodologies require full RF model support as well. As a result, the computer aided design methodology for the ESD design methods must address this issue. As an example, it may require new computer aided design methodologies that are not practiced in digital design, which are more adaptable to the RF design environment. As an example to be discussed in a later chapter, custom fixed design sizes, growable or scalable designs, parameterized cells and/or hierarchical parameterized cell ESD networks and methods of extraction for various size implementations may be required.

- RF ESD Design Chip Sub-function Synthesis: With RF ESD design, the synthesis of the digital, analog and RF segments may require unique structures in the substrate wafer or in the interconnect system, to isolate the electrical noise, and at the same time provide ESD protection between the various segments of the chip. This may require unique physical structures, and circuits to address the circuit sub-function ESD protection. Although the same ESD networks used in digital ESD design practices are utilized, the design choices are distinct due to the implications on the RF application. For example, ESD diodes can be used between ground rails between chip sub-functions. In ESD digital design, the focus may be on differential voltage isolation; for the RF ESD design practice, the focus may be on the capacitive coupling, and the impact on RF stability of networks.

- RF ESD Test Methods: In the ESD testing of RF components, unique tests need to be established on a component level and system level to evaluate the ESD degradation. Unique RF testing methods are needed which address different dc and RF parametrics degradation to evaluate the pre- and post- ESD stress test conditions. A distinction between digital ESD design practice and RF ESD design practice is the digital ESD design practice focus on dc voltage shifts, and leakage; in RF ESD design practice, the focus will be on the RF parameters and what occurs first – dc or RF degradation [8]. These methods may contain RF methods such as time domain reflection (TDR), and time domain transmission (TDT) methods.

- RF ESD Failure Criteria: In RF applications, the functional requirements are very distinct from digital applications. Unique RF parameters and ESD failure criteria need to be established based on the RF parameters, dc parameters and system level requirements. This is distinct from typical digital applications which only require dc leakage evaluation.

- RF ESD Test Systems: To address the RF ESD test methods, and failure criteria, new RF ESD test systems may be required that address product evaluation. RF ESD test systems may require ESD systems that allow extraction of the RF parameters in-situ for noise figure (NF), gain (G), output intercept third order harmonics (OIP3), as well as dc leakage evaluation. This may influence the direction of ESD human body model (HBM), machine model (MM), and transmission line pulse (TLP) systems. Today, 50 Ω-based TLP systems are compatible with 50 Ω-based RF circuits. Additionally, future TLP systems may be influenced by the needs of RF circuits.

- ESD Frequency Spectrum vs. Functional Applications: In advanced RF designs, the RF circuits are significantly faster than the ESD phenomenon; this allows for frequency "bands" for the ESD phenomenon and ESD circuit element response versus the RF functional circuit operation, and application frequency. As the RF application frequency exceeds 5 GHz, the application frequency will exceed the ESD CDM energy spectrum (e.g., approximately less than 5 GHz). As the application frequency exceeds the ESD phenomenon, the RF ESD design methodology allows for utilization of the difference in the response during ESD event time scale (e.g., frequency), and application frequency.

- ESD Frequency Domain Load Reduction Methods: In RF ESD design, a higher focus is used to lower the loading effects by taking advantage of the frequency response of the RF networks being distinct from the ESD phenomenon.

- ESD Method of Co-synthesis of ESD and RF Circuits: In the RF design of ESD networks, it is necessary to design the ESD device in conjunction with the RF circuit. By co-synthesis of the network, the loading effect as well as frequency modifications can be optimized to prevent the limitation of the ESD RF network.

- ESD Method of Utilization of ESD Element as a Capacitor in RF design: ESD elements can serve as capacitor elements. Hence, in the co-synthesis a method of transfer of the capacitance from the functional circuit to the ESD element to achieve the same RF performance is achievable.

- ESD Method of Series to Parallel Conversion of RF element to ESD element: Given a functional RF circuit which is defined as a series configuration, the representation can be modified to a parallel configuration. In the transformation from a series configuration to a parallel configuration, some portion of the element can be utilized for ESD protection using it as a parallel shunt to power or ground rail.

- ESD Method of Utilization of ESD Element as a Shunt Capacitor: Given a functional RF circuit which is defined as a series capacitor configuration, the representation can be modified to a parallel configuration in order to establish a shunt capacitor equivalent circuit. Given a capacitor in series with a resistor element, this circuit can be transformed into an equivalent circuit of a resistor and capacitor in parallel which achieves the same quality factor (Q). The transform of the network with a matched Q achieves the same circuit response. In this fashion, a series capacitor element can be substituted for a shunt ESD element which serves as an ESD element in either diodic operation or breakdown mode of operation.

- ESD Method of Parallel Susceptance Equivalent Load Compensation: Capacitive loads which are in parallel configuration can be transformed as treating two parallel susceptances. In the implementation, the total susceptance load is the parallel configuration of the new load susceptance and the ESD susceptance. Hence, the transformation of the total load susceptance to an equivalent parallel configuration of an ESD susceptance load and a new susceptance load is achieved.

- ESD Method of Series Inductive De-coupling of ESD Element Circuit: Using inductor elements in series with the ESD element, the loading effect of the ESD element can be inductively isolated. A series inductor providing a low L di/dt during ESD events allow for the current to flow through the ESD element to a power rail or ground. During functional RF operation, the L di/dt allows voltage isolation of the ESD element.

- RF ESD Method of Narrow Band Fixed Load Absorption and Resonant: Matching L-Matching Compensation Method: ESD elements can serve as a means to provide impedance matching between the output and the load. Hence, using matching techniques, the ESD element can serve as the matching elements to provide optimum matching conditions. Using a L-match circuit, consisting of a series inductor and a shunt capacitor (e.g., ESD element) the ESD network can be used as a means to provide matching between the source and the load. The shunt capacitor must remain on a constant conductance circle on a Smith admittance chart.

- RF ESD Method of Narrow Band Fixed Load ESD Absorption and Resonant Matching L-Match: For "absorption matching" the stray reactance is absorbed into the impedance matching network up to the maximum that is equal to the matching components. For "resonance matching" stray reactance is resonated out with an equal and opposite reactance, providing cancellation. Hence, the stray reactance serving as an ESD element can be resonance matched to an inductor of equal reactance. An inductor element in parallel with the ESD capacitor element can null the ESD capacitance loading effect providing "resonance matching" which hides the ESD capacitive element by matching the inductor susceptance.

- ESD Method of Cancellation: Using RF components, the loading effect of an ESD element can be hidden at the application frequency. Cancellation of the ESD loading effect can be achieved by proper loading of additional elements.

- ESD Method of Impedance Isolation: Using inductors in series with an ESD network, the inductors can serve as high impedance elements such that the loading effect of the ESD element is not observed at application frequencies.

- ESD Method of Impedance Isolation using LC tank: Using inductor and capacitor in parallel, an LC resonator tank in series with an ESD element, the loading effect of the ESD element can be reduced. The frequency of the LC tank is such that it allows operation of the ESD element, but provides isolation during functional RF operation.

- ESD Method of Lumped versus Distributed Load: In RF ESD design, ESD design focus on load reduction is achieved in the frequency domain by taking advantage of the distributed ESD loads instead of single component lumped elements. The "distributed" versus "lumped" design method can be achieved within a given element, or multiple elements.

- ESD Method of Distributed Design using Design Layout: In RF ESD design, ESD design focus on load reduction is achieved in the frequency domain by taking advantage of the distributed nature of a single ESD element. This can be achieved through design layout by introducing resistance, capacitance or inductance within a given ESD design layout. Metal interconnect design and layout distribution within diodes, MOSFETs, and bipolar transistors can introduce distributed effects. Where this is typically undesirable in a digital operation of the ESD elements, in an RF application, it can be intentionally utilized.

- ESD Methods of Distributed Design using Multiple Circuit Element Stages: In RF ESD design, ESD design focus on load reduction is achieved in the frequency domain by taking advantage of multiple elements. This can be achieved by multiple stage designs of equal, or variable size stages with the introduction of resistance, capacitance or inductance within a given ESD multiple-stage design. This can be achieved through introduction of RF resistor, capacitor or inductor components into the ESD implementation; whereas this is typically undesirable in a digital operation of the ESD elements, in an RF application, this can be intentionally utilized.

- ESD Method of Resistive De-coupling Using Distributed Multiple Circuit Element Stages: In RF ESD design, ESD design focus on load reduction is achieved in the frequency domain by taking advantage of multiple elements and series resistors. This can be achieved by multiple stage designs of equal, or variable size stages with the introduction of resistors within a given ESD multiple-stage design. The introduction of resistors provides an IR voltage drop isolating the successive stages during functional operation, but not during ESD operation.

- ESD Method of Inductive De-coupling Using Distributed Multiple Circuit Element Stages: In RF ESD design, ESD design focus on load reduction is achieved in the frequency domain by taking advantage of multiple elements and inductors. This can be achieved by multiple stage designs of equal, or variable size stages with the introduction of on-chip or off-chip inductors within a given ESD multiple-stage design. The introduction of inductors produces a L di/dt voltage drop isolating the successive stages during functional operation, but not during ESD operation.

- ESD Methods of Distributed Design using Co-planar Waveguides: In RF ESD design, multi-stage implementations can place co-planar waveguides (CPW) to provide improvements in the power transfer, matching, and reduce the loading effect on the input nodes.

- ESD Method of Distributed Design for Digital Semiconductor Chip Cores: Digital chip sectors are not inherently designed for 50 Ω matching conditions. Hence, an RF design practice is to place a resistive element shunt for utilization of distributed design for core chip sub-functions.

- ESD Method of Capacitive Isolation Buffering: Using de-coupling capacitors in series with RF elements can provide impedance buffering of receiver networks, allowing operation of ESD networks. Capacitor elements can be metal-insulator-metal (MIM) capacitors, vertical parallel plate (VPP) capacitors, or metal- inter-level dielectric layer – metal (M/ILD/ILD) capacitors. This method can not be utilized in dc circuits due to the blocking of dc currents.

- ESD Method of Architecture and ESD circuit for Improved Linearity: ESD networks can be designed in a fashion to eliminate linearity issues in RF design. For example, diode elements and varactor structures have capacitance variation as a function of applied voltage. Using ESD elements (e.g., double diode configuration), RF circuit linearity can be improved.

- RF ESD Method of Tuning an ESD circuit for Improved Linearity: ESD networks can be designed in a fashion to improve linearity issues in RF design by tuning. For example, diode elements and varactor structures have capacitance variation as a function of applied voltage. These variations can be modified by variable tuning by using tunable ESD elements with semiconductor process or design layout techniques.

- RF ESD Method of Noise and ESD Optimization: Noise is a concern in RF circuits. This influences the chip architecture between the digital, analog, and RF circuits. Additionally, noise concerns also determine the substrate doping concentration, semiconductor profile, isolation strategy and guard ring design. Additionally, noise may determine the acceptable ESD device type due to noise concerns. Hence, the method of co-synthesis of chip architecture, chip power grids, choice of ESD elements, and choice of ESD circuits are all influenced by the noise requirements.

- RF ESD Method of Quality Factor and ESD Optimization: The quality factor, Q, is influenced by the ESD device choice. Additionally, Q degradation can occur in RF passive elements such as resistors, inductors and capacitors from ESD events. Hence Q optimization and the ESD current path optimization are needed to have RF degradation mechanisms associated with ESD stress of critical circuit elements.

- RF ESD Method of Stability and ESD Optimization: In RF ESD design, circuits must be designed to achieve electrical dc stability, thermal stability, and RF stability. Amplifier stability is a function of the stability at both the source and load levels. In RF design, these are defined as source and load stability circles. The stability of the source and load is a function of the minimum resistance requirement. With the addition of ESD ballast resistance, circuit stability can be improved. Co-synthesizing the stability requirement, ESD resistance can be integrated to improve circuit stability.

- RF ESD Method of Gain Stability, Noise, Q, and ESD Optimization: In the optimization of circuits, the gain stability, noise, quality factor and the ESD can be co-synthesized. An ESD circuit can be designed such that the ESD elements are added to a circuit to help satisfy the Stern stability criteria. For ESD optimization, the path from source to load along the gain-noise optimum contour which has the maximum shunt capacitance will achieve this optimized solution.

- ESD Circuits which are Non-frequency Triggered: The introduction of ESD circuits with frequency-initiated trigger elements, such as RC-triggered MOSFET ESD power clamps, or RC-triggered ESD input networks, can be undesirable due to interaction with other RF circuit responses. For example, the introduction of RC-triggered networks that have inductor loads can introduce undesired oscillation states, and functional issues. Hence, in RF technology, non-frequency initiated trigger networks are desirable for some RF applications (e.g., voltage-triggered networks).

- ESD System Level and Chip Level Multi-stage Solutions: At radio frequency application frequencies, ESD protection loading effects have significant impact on the RF performance. ESD solutions for RF application include the combination of both on-chip and off-chip ESD solutions: spark gaps, field emission devices (FED), transient voltage suppression (TVS) devices, polymer voltage suppression (PVS) devices, mechanical shunts, and other solutions. By combining both off- and on-chip ESD protection solutions, the amount or percentage of on-chip protection solutions can be reduced.

- ESD Non-semiconductor Devices: Spark gaps, field emission devices (FED), transient voltage suppression (TVS) devices, polymer voltage suppression (PVS) devices, mechanical package "crowbar" shunts, and other solutions are utilized in RF applications off-chip due to loading effects, space (e.g., ESD design area), cost (e.g., cost/die), or the lack of the proper material to form ESD protection circuitry (e.g., substrate material). These solutions are not typically an option in semiconductor chips with high pin count and packaging constraints, but for low pin-count low circuit density applications, these are an option.

5.5.2 ESD RF Circuits – Signal Pin ESD Networks

RF receiver circuits are very important in RF ESD design because of the ESD sensitivity of these networks. Typically, the receiver circuits are the most sensitive circuits in a chip application. Receiver performance has a critical role in the semiconductor chip performance. The primary reasons for this are as follows [4,5,7,13]:

- RF receiver circuits are small in physical area.

- Receiver performance requirements limit the ESD loading allowed on the receiver. MOSFET gate area, bipolar emitter area, and electrical interconnect wiring widths impact the receiver performance.

- RF receiver input are electrically connected to either the MOSFET gate (in a CMOS receiver) where the MOSFET gate dielectric region is the most ESD sensitive region in RF MOSFET receiver networks. RF MOSFET gate dielectric scales with the RF performance objectives.

- RF receiver input are electrically connected to the bipolar base region (in a bipolar receiver) where the bipolar transistor emitter-base junction is the most ESD sensitive region of the bipolar transistor. The base region scales with RF performance objectives.

- Both the MOSFET gate dielectric region and the bipolar transistor base region are the more sensitive regions of the structures.

- RF receivers require low series resistance.

Receiver circuits are a common ESD sensitive circuit in bipolar and BiCMOS technology. Bipolar receiver circuits typically consist of npn bipolar transistors configured in a common emitter configuration. For bipolar receivers, the input pad is electrically connected to the base contact of the npn transistor, with the collector connected to V_{CC} either directly or through additional circuitry. The npn bipolar transistor emitter is electrically connected to V_{SS}, or through an emitter resistor element, or additional circuitry.

In bipolar receiver networks, for a positive polarity HBM ESD event, as the base voltage increases, the base-to-emitter voltage increases, leading to forward biasing of the base-emitter junction. The base-emitter junction becomes forward active, leading to current flowing from the base to the emitter region. Typically in bipolar receiver networks, the physical size of the emitter regions is small. When the ESD current exceeds the safe operation area (SOA), degradation effects occur in the bipolar transistor. The bipolar device degradation is observed as a change in the transconductance of the bipolar transistor. From the electrical parametrics, the unity current cutoff frequency, f_T, decreases with increased ESD current levels. From a f_T–I_C plot, the f_T magnitude decreases with ESD pulse events, leading to a decrease in the peak f_T.

For a negative pulse event, the base-emitter region is reverse biased. As the voltage on the signal pad decreases, the base-emitter reverse-bias voltage across the base-emitter metallurgical junction increases. Avalanche breakdown occurs in the emitter-base metallurgical junction, leading to an increase in the current flowing through the emitter and base regions; this leads to thermal runaway and bipolar second breakdown in the bipolar transistor. The experimental results show that the negative polarity failure level has a lower magnitude compared to the positive polarity failure level.

One common ESD design solution used to provide improved ESD results in a single-ended bipolar receiver network is to place a p-n diode element in parallel with the npn bipolar transistor emitter-base junction (Figure 5.19). Using a parallel element, the p-n junction is placed such that the anode is electrically connected to the npn emitter, and the cathode is electrically connected to the npn base region; this ESD element serves as a bypass element avoiding avalanche breakdown of the npn base-emitter junction. The diode element is placed local to the npn transistor element to avoid substrate resistance from preventing early turn-on of the ESD diode element. Note that this element is analogous to the CDM solution used in CMOS receiver networks. For a bipolar transistor, it serves events from both the signal pad and potentially the emitter electrode.

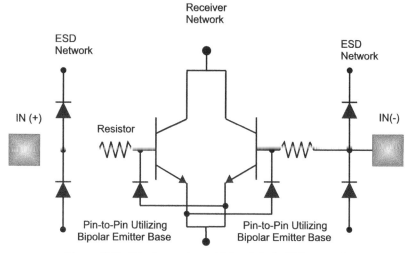

Figure 5.19   Bipolar receiver with local diode ESD network

In radio frequency (RF) bipolar receivers, metal-insulator-metal (MIM) capacitors are used between the signal pad and the base electrode. For positive or negative mode polarity events, the MIM capacitor can fail due to dielectric degradation. Without ESD protection on the receiver network, the ESD failure levels of the receiver network will be limited by the MIM capacitor element. An ESD solution to prevent ESD failure in these RF bipolar receivers is to use a p-n diode element in parallel with the MIM capacitor element. The p-n diode element can be in a reverse configuration so that it serves as a parallel capacitor element, and does not allow a dc voltage to be transmitted between the signal pad and the bipolar receiver base element. The functional disadvantage of the p-n element is the impact of the effective quality factor "Q" of the capacitor element.

For differential pair receiver networks, the concepts of common centroid design, and signal pin-to-signal pin protection can be applied to RF networks. The solution will vary dependent on whether it is a CMOS differential pair receiver, or a bipolar differential pair circuit.

5.5.3 ESD RF Circuits – ESD Power Clamps

For RF BiCMOS technology, ESD power clamps can be constructed from both the digital, analog, or RF power domains. For the low voltage CMOS power domains, GGNMOS or RC-triggered ESD MOSFET power clamps can be used. For the bipolar section of the semiconductor chip, bipolar ESD power clamps are preferable. An example of an ESD bipolar power clamp is a circuit which uses a voltage triggered bipolar transistor, and a bipolar "clamp" element [4,5,7,13].

With this ESD bipolar power clamp, fundamental theory associated with the power and performance of a bipolar transistor applies (Figures 5.20 and 5.21). A bipolar-based ESD power clamp that utilizes the breakdown of a bipolar transistor in a collector-to-emitter configuration can be synthesized using a first transistor for the trigger element

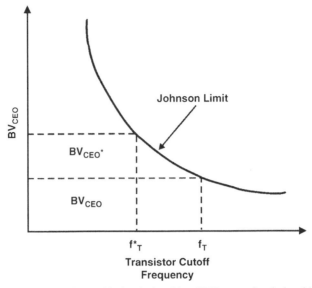

Figure 5.20 Johnson Limit relationship of BV_{CEO} vs f_T relationship

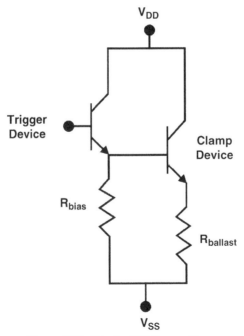

Figure 5.21 Bipolar ESD Power Clamp

and a second transistor as the output clamp device, as first shown in Figure 5.4. A BV_{CEO} voltage-triggered bipolar ESD power clamp contains an output bipolar transistor between the first and second power rail where the first power rail is electrically connected to the bipolar transistor collector, and the second power rail is electrically connected to the bipolar transistor emitter. A bias resistor element is electrically connected to the base of the output clamp device. The bias resistor sets the base to a low potential to prevent the "turn-on" of the output clamp [4,5,13].

Bipolar BV_{CEO} breakdown voltage triggered ESD power clamp in bipolar and BiCMOS technology has ESD advantages as follows:

- Low trigger voltage conditions.
- Scalable.
- Compatibility and design integration with analog and radio frequency (RF) circuits.
- Compatibility with bipolar transistors.
- Use of supported bipolar transistor (e.g., non-use of parasitic devices).
- Circuit simulation.
- Scalable.
- Utilization of multiple transistors.
- Low noise source.
- Use for positive or negative polarity power supplies.

The BV_{CEO} breakdown voltage triggered ESD power clamp can utilize a transistor in a common-emitter mode and initiate the output clamp at this voltage condition. A unique aspect of this implementation is by using the BV_{CEO} condition when there is an inherent inter-relation with the unity current cutoff frequency of the transistor. From the Johnson Limit relationship, its power formulation is given as

$$(P_m X_c)^{1/2} f_T = E_m v_s \Big/ {2\pi}$$

where P_m is the maximum power, X_C is the reactance $X_c = \frac{1}{2} \Pi\, f_T\, C_{bc}$, f_T is the unity current gain cutoff frequency, E_m is the maximum electric field and v_s is the electron saturation velocity. Expressed as the product of the maximum voltage, V_m, and the cutoff frequency,

$$V_m f_T = E_m v_s \Big/ {2\pi}$$

Hence from the Johnson Limit equation,

$$V_m^* f_T^* = V_m f_T = E_m v_s \Big/ {2\pi}$$

where $V_m^* f_T^*$ is associated with a first transistor and $V_m f_T$ is associated with a second transistor. The ratio of breakdown voltages can be determined as

$$\frac{V_m^*}{V_m} = \frac{f_T}{f_T^*}$$

Using this Johnson relationship, an ESD power clamp can be synthesized where a trigger device with the lowest breakdown voltage can be created by using the highest cutoff frequency (f_T) transistor and a clamp device with the highest breakdown device having the lowest cutoff frequency (f_T).

A BV_{CEO} breakdown voltage-triggered bipolar power clamp can be synthesized from this relationship between the power supplies. In this configuration, ESD power clamp is in a common-collector configuration. For this configuration to be suitable as an ESD power clamp, we can take advantage of the inverse relationship between the BV_{CEO} breakdown voltage and the unity current gain cutoff frequency, f_T, of the device. For an ESD power clamp, the ESD output clamp device must have a high breakdown voltage in order to address the functional potential between the V_{CC} power supply and ground potential. This ESD power clamp requires an f_T value above the ESD pulse frequency to discharge the current effectively. For the bipolar trigger device, a low BV_{CEO} breakdown voltage device is needed in order to initiate base current into the clamp device at an early enough voltage (Figure 5.21).

This circuit can be constructed in a homo-junction silicon bipolar junction transistor (BJT), or a silicon germanium, silicon germanium carbon, or gallium arsenide hetero-junction bipolar transistor (HBT). The bipolar-based BV_{CEO} triggered ESD power clamp trigger network consists of a high f_T SiGe HBT with a bias resistor. When the transistor collector-to-emitter voltage is below the breakdown voltage, no current is flowing through the trigger transistor. The bias resistor holds the base of the SiGe HBT clamp transistor to a ground potential. With no current flowing, the output clamp can be visualized as a "grounded base" npn device between the power supplies. When the voltage on V_{CC} exceeds the collector-to-emitter breakdown voltage, BV_{CEO}, in the high f_T SiGe HBT, current flows into the base of the SiGe HBT high breakdown device. This leads to discharging of the current on the V_{CC} (or V_{DD}) power rail to the V_{SS} ground power rail.

5.5.4 ESD RF Circuits – ESD RF VSS-to-VSS Networks

ESD protection networks are also placed between the digital and RF ground rails. For RF implementations, the introduction of RF V_{SS}-to-V_{SS} ESD networks can lead to RF degradation if the capacitive coupling between the ground rails is excessive. As in digital implementations, anti-parallel diode strings can be added between the grounds. For a digital implementation, the number of diodes in series is associated with the desired differential voltage required for noise isolation; for RF implementations, the number of diodes is dependent on the capacitive de-coupling needs for the RF circuits.

5.6 SUMMARY AND CLOSING COMMENTS

In Chapter 5, ESD protection networks are discussed as the cure to the problem of ESD concerns on semiconductor components. Examples of ESD input circuits and ESD power clamps are given as examples of typical solutions used from 1 μm technology to 28 nm technology. The chapter provided a brief discussion for ESD on-chip concepts for digital, analog and RF circuitry. In the 1990s, there were few books on the subject, but today for readers who desire detailed information, there is a vast amount of books and publications today for your perusal.

In Chapter 6, we shift to the system level solutions for ESD. In the next chapter, the issue of non-traditional ESD on-chip and off-chip protection networks will be discussed as the solutions for system-level ESD problems. Spark gaps, field emission devices and conductive polymer ESD protection concepts are a few of the examples. New testing methodologies are discussed to evaluate system level response to component level ESD degradation, as well as sensitivity to electromagnetic fields. Two examples are shown, where the "degraded component" was placed into the system to evaluate the system level performance degradation. Additionally, an ESD/EMC scanning system was shown that addresses both component and system level susceptibility. This new methodology bridges the gap between component level ESD evaluation and EMC system level testing. Also, this ESD/EMC test method provides a mapping of the locations of susceptibility, finding the root cause, and allowing for re-design or improvements.

REFERENCES

1. Voldman, S. (2004) *ESD: Physics and Devices*, John Wiley and Sons, Ltd., Chichester, England.
2. Voldman, S. (2005) *ESD: Circuits and Devices*, John Wiley and Sons, Ltd., Chichester, England.
3. Voldman, S. (2008) *ESD: Circuits and Devices*, Publishing House of Electronic Industry (PHEI), Beijing, China.
4. Voldman, S. (2006) *ESD: RF Circuits and Technology*, John Wiley and Sons, Ltd., Chichester, England.
5. Voldman, S. (2011) *ESD: RF Circuits and Technology*, Publishing House of Electronic Industry (PHEI), Beijing, China.
6. Voldman, S. (2009) *ESD: Failure Mechanisms and Models*, John Wiley and Sons, Ltd., Chichester, England.
7. Voldman, S. (2005) *ESD: Design and Synthesis*, John Wiley and Sons, Ltd., Chichester, England.
8. Dabral, S. and Maloney, T.J. (1998) *Basic ESD and I/O Design*, John Wiley and Sons Ltd., West Sussex.
9. Wang, A.Z.H. (2002) *On Chip ESD Protection for Integrated Circuits*, Kluwer Publications, New York.
10. Amerasekera, A. and Duvvury, C. (2002) *ESD in Silicon Integrated Circuits*, 2nd edn, John Wiley and Sons, Ltd., West Sussex.
11. Hastings, A. (2006) *The Art of Analog Layout*, 2nd edn, Pearson Prentice Hall, New Jersey.
12. Vashchenko, V. and Shibkov, A. (2010) *ESD Design in Analog Circuits*, Springer, New York.
13. Singh, R., Harame, D., and Oprysko, M. (2004) *Silicon Germanium: Technology, Modeling and Design*, John Wiley and Sons.

6 ESD in Systems – Problems and Solutions

I was in Palenque, Mexico in 2011, in a café drinking some good coffee from the province of Chiapas, before we visited the Mayan ruins. In this peaceful town, we then heard the clanging of metal cans, and pots which were banging on the ground by metal wires off the back of the truck . . . This noisy open truck was filled with propane tanks, and a few workers in the back of the truck. At first, I thought it was a celebration, a wedding? . . . No, it just was an ESD concern . . . and a noisy system solution!

6.1 ESD SYSTEM SOLUTIONS FROM LARGEST TO SMALLEST

In this chapter, solutions for problems will be discussed in systems from airplanes, cars, computers, disk drives to semiconductor chips. The chapter will discuss system level "off-chip" ESD protection such as spark gaps [1–6], field emission devices (FED) [7], transient voltage suppression devices [8,9], mechanical package solutions [10–22], to magnetic recording in-line and package solutions [23–27]. The discussion will highlight these system level solutions and their relationships to ESD, EMI, and EMC [28–40]. The chapter will continue with discussions of IEC issues [41–45], on-chip architecture and floor planning in digital, analog and RF design [46–54], and close with discussions on new ESD/EMC scanning techniques.

6.2 AEROSPACE SOLUTIONS

When you get on an airplane, ever notice they refuel when the passengers are off the plane? Ever notice the gas truck has an electrical grounding wire with a metal clip to

ESD Basics: From Semiconductor Manufacturing to Product Use, First Edition. Steven H. Voldman.
© 2012 John Wiley & Sons, Ltd. Published 2012 by John Wiley & Sons, Ltd.

attach to the plane? The airline business must follow procedures every day to address charging in airplanes. Additionally, standards and procedures exist to address the environmental conditions (e.g., RTCA/DO-160E. *Environmental Conditions and Test Procedures for Airborne Equipment*, Radio Technical Commission for Aeronautics (RTCA), December 7, 2004) [35].

6.3 OIL TANKER SOLUTIONS

> *. . . On the Atlantic Coasts, oil tankers would come near the shore, and a pipe would be connected to pump the oil off the tanker to the round tank storage facilities. As the oil was being pumped from the tanker in the water, the oil tank storage facilities would explode.*

When this problem first occurred, it was realized that the charged ions in the oil were charging up the oil storage tank. Without the correct grounding of the pipe between the oil tanker, and the oil storage facility, the moving charge was not grounded to the ground potential. The solution was found that by grounding the pipe along the path, the electrical charge would move to the sidewalls of the pipe, and be electrically grounded.

Today, there is some belief that static electricity is still a problem in some of the recent tanker explosions. Events occur around the globe of fires on tankers, and there is potentially electrostatics involved.

6.4 AUTOMOBILE SOLUTIONS

> *. . . It is 66 degrees in Burlington, Vermont so I went jogging – when I went past the gas station, I noticed fire extinguishers are being placed next to the pumps!! ESD, it is everywhere!*

In recent years, a solution for automobile and gas station related explosions and fires is being addressed by the following solutions:

- ESD Awareness.

- ESD Procedures – Do's and Dont's.

- Fire Extinguishers.

On the gas station pumps, I noticed it gave warnings of portable containers (aka gas cans). It stated that portable containers are not to be placed on mats, or the back of trucks when filling up the gas can. The correct procedure is to place the gas can on the ground during the filling process.

From this discussion, today it is still an issue for airplanes, oil tankers to automobiles . . . moving charged fluid must have the correct grounding and procedures in place to avoid ESD-related fires, and explosions.

6.5 COMPUTERS – SERVERS

Servers historically were large and expensive. Cable installation was a concern. Solutions to avoid cable discharge events into systems include "touch pads," handling procedures, surge protection devices, and on-chip protection.

6.5.1 Servers – Touch Pads and Handling Procedures

On large servers, there are processes and procedures for power up, power down, and cable installation. One of the procedures for installation of cables was to use a "touch pad." The operator was to touch the "touch pad" while holding a cable, and then after discharging to the chassis of the server, plug the cable into the system.

6.6 MOTHER BOARDS AND CARDS

ESD, EOS, EMC and EMI solutions are integrated into cards and boards to avoid failures in systems [33]. Solutions include some of the following:

- System Card Insertion Contacts.
- System Level Ground Design.
- System Level "On Board" EOS and ESD Protection.

6.6.1 System Card Insertion Contacts

A commonly used technique to avoid system failures is to design card contacts so that the ground (V_{SS}) and power (V_{DD}) are extended beyond the signal pins. The contacts of the power and ground are designed such that as the card is inserted into a system socket, contact occurs first with power and ground (Figure 6.1). The design is such that if the board (or card) was charged, the charge would flow to the system power or system ground first. Historically, there was a hand held device which did not design this correctly. When the hand held device was inserted into its socket to interface with a computer, a current pulse went into the computer signal lines, leading to computer failures.

6.6.2 System Level Board Design – Ground Design

Ground design is very critical in system board design [28–40]. In many board designs, the board designer separates the digital ground from the analog ground for system level noise isolation. Signal lines on the board that cross the two ground planes are vulnerable to electromagnetic interference on the signal lines.

A solution to address EMI and EMC problems is to avoid separation of the ground planes but have them connected at some location within the board [33]. Figure 6.2 shows an

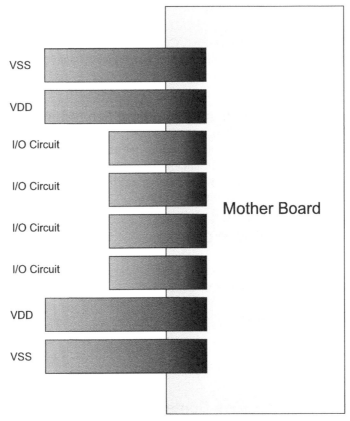

Figure 6.1 Contact design with extended V_{DD} and V_{SS}

example of a digital and analog application, where the digital circuitry is in one section of the ground plane, and the analog circuitry is in the other section of the same ground plane. The ground plane has a small "bridge" where the digital to analog converter is placed. In this fashion, all the digital signal lines and pins remain on the one side of the plane; this avoids

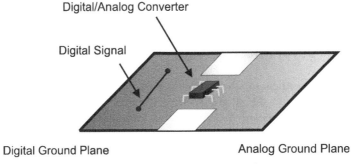

Figure 6.2 Single ground board design

digital noise from impacting the analog circuitry but at the same time does not have a fully separated ground plane.

6.7 SYSTEM LEVEL "ON BOARD" ESD PROTECTION

For semiconductor components, ESD protection is typically referred to as "on-chip" ESD protection. "On board" ESD protection is ESD components placed on the card or board; this can be instead of on-chip ESD protection, or additional protection. In the case where the "on board" ESD protection is the only source of ESD protection for the system, there are reasons why this is done:

• Cost.

• Chip Area.

• Chip Size.

• On-chip is not possible.

Applications that do not use "on-chip" ESD protection include magnetic recording industry, single component lasers, light emitting diodes, ultrahigh voltage devices, and some RF technologies (e.g., Gallium Arsenide, Gallium Nitride). In some of these cases it is the incompatibility of the wafer with the device – such as the magnetic recording devices. In some cases, on-chip devices are not used due to area – such as lasers, and RF technologies [49,50].

With increasing RF performance objectives, the ability to provide low capacitance ESD protection will increase in importance. As the application frequency increased to 1 GHz, the need to reduce the size of the ESD protection loading effects became an issue for CMOS, silicon-on-insulator (SOI), and BiCMOS technology. As the application frequencies increased from 1 to 5 GHz, the choices of the ESD element based on its ESD robustness versus capacitance loading is evaluated to provide the optimum ESD solution. As the frequency increased above 5 GHz, co-synthesis of ESD and RF performance increased in interest. The question remains, what will be the ESD chip and system solution as the frequency of the circuits increases to 10 and, 100 GHz, and above. Today, 100 GHz circuits have been demonstrated.

In high speed RF applications, traditional solutions are abandoned, and new ESD directions will be used. The chapter will focus on air breakdown spark gaps [1–7], field emission devices (FED) [7], polymer voltage suppression (PVS) devices [8,9], and mechanical packaging solutions [10–28].

On-board ESD protection uses different components for solutions. On-board ESD protection includes the following:

• Spark gaps.

• Single component diode elements.

• Single component capacitors.

• Transient voltage suppression (TVS) devices.

6.7.1 Spark Gaps

With increasing RF performance objectives, the ability to provide low capacitance ESD protection will increase in importance. Spark gaps may play an important role in on-chip or off-chip ESD protection in the future. Spark gaps perform on the concept of air breakdown [1–7]. Spark gaps have been of interest in the systems for circuit boards and on modules in the 1970s [5,6]. Spark gaps have been formed on the printed circuit boards, modules and multi-chip environments. Spark gaps can be formed on ceramic substrates, silicon carriers and other forms of packages that semiconductor chips are mounted upon. Off-chip spark gaps have the advantage of not using areas on a semiconductor chip.

Spark gaps and field emission devices (FEDs) can be constructed on-chip or off-chip as a form of ESD protection [5–7]. Spark gaps can be formed using metallization patterns formed on the package or substrate material using closely spaced metal lines. Spark gaps formed on the ceramic substrate are limited by the allowed metal line-to-line spacing. A disadvantage of the off-chip implementation of a spark gap is the line-to-line spacing being significantly larger than what is achievable in an on-chip implementation. Spark gap reliability is critical; this manifests itself in the ability to discharge, as well as repeatability being an important issue in using spark gap as ESD protection. Another limitation of spark gaps is the reaction time. Gas discharges which occur in the spark gaps have a nanosecond reaction time. Another limitation is the breakdown voltages that occur. Hence, spark gaps have the following limitations [7]:

- Reliability of electrical discharge initiation.

- Repeatability of discharge current and voltage magnitude.

- Time constant response of the arc discharge.

- Magnitude of the breakdown voltages (e.g., compared to the application device sensitivity).

During discharge events, electrical damage can occur to the spark gap electrodes leading to reliability concerns. Secondly, with damage to the electrode curvature, and material residuals, the repeatability may be inadequate to insure or qualify the off-chip spark gap.

Additionally, for RF applications, the time constant associated with the breakdown, gas ionization, and the reverse recovery time of the ionized gas may be long compared to circuit response of RF circuity. Spark gap initiation time is a function of the collision and ionization times; the time constant to initiate ionization of gas molecules associated with avalanche multiplication [1–4]. Breakdown in gases is initiated by a feedback induced by the acceleration of carriers leading to secondary carriers. At very high speeds, the ability to provide semiconductor devices may be limited.

And lastly, another concern is the magnitude of the trigger voltage of the spark gap. Paschen studied the breakdown physics of gases in planar gap regions [1]. The result of Paschen showed that the breakdown process is a function of the product of the gas pressure and the distance between the electrodes. As discussed in Chapter 1, Paschen showed that

$$pd \approx \frac{d}{l}$$

where p is the pressure, d is the distance between the plates and l is the mean free path of the electrons. From the work of Paschen, a universal curve was established which followed the same characteristics independent on the gas in the gap. The Paschen curve is a plot of the logarithm of the breakdown voltage as a function of the logarithm of the product of the pressure and gap distance.

$$V_{BD} = f(pd)$$

At very low values of the p-d product, electrons must accelerate beyond the ionization limit to produce an avalanche process because the likelihood of impacts is too low. In this region, the breakdown voltage decreases with the increasing value of the pressure-gap product. This occurs until a minimum condition is reached. At very high values of the pressure-gap product, the number of inelastic collisions is higher and the breakdown voltage increases. This U-shaped dependence is characteristic of the gas phenomenon. Townsend, in 1915, noted that the breakdown occurs at a critical avalanche height [4],

$$H = e^{\alpha d} = \frac{1}{\gamma}$$

In this expression, the avalanche height H, is equal to the exponential of the product of the probability coefficient of ionization (number of ionizing impacts per electron and unit distance in the direction of the electric field) and electrode spacing. The avalanche height, H can also be expressed as the inverse of the probability coefficient of regeneration (number of new electrons released from the cathode per positive ion).

These devices may also be limited by the resistance of the arc. As discussed in Chapter 1, Toepler, in 1906, established a relationship of the arc resistance in a discharge process [2]. Toepler's law states that the arc resistance at any time is inversely proportional to the charge which has flowed through the arc

$$R(t) = \frac{k_T D}{\int_0^t I(t')dt'}$$

where I(t) is the current in the arc discharge at time t, and D is the gap between the electrodes. The value k_T is a constant whose value is 4×10^{-5} V-sec/cm.

For off-chip spark gaps, the ability to produce a well-defined metal spark gap is difficult due to the lithography and semiconductor process techniques used in the packaging environments. Experimental work on off-chip spark gaps had limited success for HBM protection. Due to the nature of the spark gap breakdown, it was found that "ESD test windows" were observed. Using commercial human body model (HBM) test systems, it was found that off-chip spark gap devices could provide ESD protection at high voltages (e.g., 1000 to 2000 V HBM). But, with an incremental ESD step-stress at 100 V increments, ESD failure occurred at lower HBM stress levels. Where it was found that the HBM stress provided protection for higher voltages, semiconductor products were vulnerable at lower HBM levels. Hence, it is important in the application of ESD spark gap elements to demonstrate the operation window, and its operability range.

Experimental work on off-chip spark gaps were also found to have had limited success in improving ESD protection. H. Hyatt utilized a three-stage system protection strategy: a first system level surge protection element, a second-stage on-package spark gap, and third, on-chip ESD protection network used to protect a monolithic microwave semiconductor chip. A very-fast transmission line pulse (VF-TLP) test source, with a 1 ns pulse width demonstrated that the second stage off-chip spark gap (e.g., integrated into the package module) showed no improvement in the three-stage ESD system. The off-chip spark gap had limited response as a result of the lithographic capability to construct on a package (e.g., spark gap width was too large), and limited reaction time for very fast transmission line pulse events.

In the experimental work of Bock, on-chip spark gaps were designed for applications of ESD protection of monolithic microwave integrated circuits (MMIC). Bock developed on-chip spark gaps with a gap dimension of 0.4 to 4.0 μm. At a 1 μm gap dimension, the breakdown voltage of air under normal pressure is in the order of 250 V. It was noted by Bock, that in the experimental work, the spark gap breakdown voltage is reduced to 45 V due to electric field enhancement factors due to non-planar surfaces [7]. The primary mode of discharge was associated with field emission in the spark gap, not the actual gas discharge phenomenon. It was noted by Bock that for on-chip spark gaps in the micron dimension, the mode of operation was partly ionization processes as well as field emission between the two electrodes [7].

The usage of spark gaps as an off-chip protection solution will be more feasible only as the physical dimensions on silicon carriers, ceramic substrates and other packages are reduced allowing a lower voltage spark gap. To make it suitable for low voltage semiconductors, optimization of the field emission properties of the spark gaps must be controlled. Additionally, advanced processing techniques will be required to lower the breakdown voltage. Additionally, as the need for off-chip ESD solutions increases, the need for improved spark gaps with improved reliability will be needed.

6.7.2 Field Emission Devices (FED)

Field emission devices (FED) as electrostatic discharge protection have advantages over silicon based devices. The advantages are as follows [7]:

- Low dielectric constant ($\varepsilon_r = 1$).
- Power dissipation.
- Power handling capability.

Compared to silicon-based ESD networks, the dielectric constant of air is unity. In semiconductor materials (e.g., silicon, gallium arsenide, germanium, silicon dioxide, low-k inter-level dielectrics), the dielectric constant is greater than unity, leading to a capacitance advantage in air-based ESD protection networks.

In the air gap itself, the power dissipation and self-heating of the device is not as significant as occurs in silicon, germanium or gallium based semiconductor devices. Hence, there the ability to handle pulsed power is significant.

From a RF performance and functionality perspective, FED ESD devices have the following advantages [7]:

- Low parasitic capacitance (e.g., less than 0.1 pF).

- Fast switching times (e.g., less than 1 ps).

- High current densities ($J > 10^8$ A/cm^2).

As the application frequency increases for MMIC applications, under a constraint of a constant reactance, the capacitance of the ESD protection will be scaled. At 100 MHz, 1 to 10 pF of capacitance was acceptable loading for ESD protection networks. At 1 GHz, 1 pF of ESD capacitance was acceptable for most CMOS digital logic and RF applications. But, at 10 GHz, and above, the ESD capacitance loading must be reduced below 0.1 pF capacitance. In semiconductor-based ESD elements this will be achievable by doping optimization or reduction of the ESD network size. Hence, the FED element can achieve the loading capacitance requirement for greater than 10 GHz frequencies.

For switching times, field emission devices (FED) are limited by the transit time of carrier. As opposed to spark gaps, field emission devices are not limited to the ionization processes and avalanche processes which occur during air breakdown. As a result, for scaled devices, the potential speed of FEDs can achieve 10^{-12} sec time scales. For ESD events, human body model and machine model specifications, the time constant is 10^{-9} sec time scales. From the work of J. Barth, the responses can range to 100×10^{-12} sec. For charged device models, the rise time is less than 250 ps time. Hence, the field emission device response time can be as fast, or faster than the time scales associated with ESD events.

Field emission devices (FED) require closely spaced structures and the ability to form a "point" electrode with a well-defined radius of curvature and emitter-to-collector spacing. Both field emission devices and spark gaps utilize air bridge structures to form the physical gap. This requires additional process masks, as well as a process where air exposure is not a reliability issue. K. Bock and H.L. Hartnagel reported field emission devices can provide higher current densities and have demonstrated higher reliability of discharge repeatability [7]. Using an additional five masks, an air-bridge GaAs field emission device (FED) structure was formed on-wafer with a loading capacitance of less than 0.1 pF [7].

In the structure, the field emission device was constructed in a GaAs process. The substrate material is a semi-insulating GaAs substrate. An n-epitaxial Gallium Arsenide layer is formed on the semi-insulating GaAs substrate. The emitter structure is formed using a wet-etch process from the n-epitaxial GaAs layer. The etch process forms the emitter points and defines the radius of curvature of the emitter structure; the radius of curvature varies the electric field at the emitter points. The emitter structure is then coated with a low work-function metal film. Using the wet-etch process, the emitter tips were designed to 75 nm dimensions [7]. Ohmic contacts are formed at or near the n-epitaxial region between the emitter points in the emitter structure of the field emission device.

To provide a well-defined emitter gap, a photo-resist film is deposited over the emitter structure, followed by a second deposition of a conformal metal film. The metal film is masked and etched to form the collector structure. After removal of the photo-resist, an air-gap is formed between the GaAs emitters and the metal air-bridge collector structure

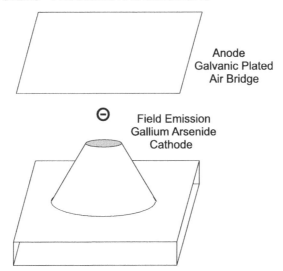

Anode
Galvanic Plated
Air Bridge

Field Emission
Gallium Arsenide
Cathode

Figure 6.3 Cross section of a field emission device

(Figure 6.3). Hence control of the design is a function of the ability to form the GaAs emitter points as well as control of the photo-resist film thickness.

Bock and Hartnagel discuss a means of optimization of called "electronic blunting effect [7]." In a semiconductor film, a saturation effect occurs which leads to a current-limiting phenomenon. As the space charge in the gap increases, the electric field penetrates into the semiconductor emitter structure, where the penetration is a function of the doping concentration. When this occurs, the series resistance of the emitter structure increases; the series resistance increase of the emitter leads to a spatial self-ballasting near the emitter point. As a result, the current distributes more evenly through the emitter region, lowering the peak current density at the emitter tip, and provides better current distribution through the emitter structure. Bock and Hartnagel point out that this "electronic blunting effect" can be utilized to provide a ballasting effect in a multi-emitter field effect device. It was also noted that in the case where a metal film is used in the field emission device, where there are non-uniform variations in the emitter height, a single emitter may turn on prior to other parallel emitter structures. But in the case of a semiconductor emitter structure, the "electronic blunting effect" can lead to simultaneous turn-on of multiple emitter structures, leading to an improved ESD protection device. Hence, although the metal film is superior from a current density perspective, for utilization for an FED ESD element, it is advantageous to use a semiconductor that leads to better current distribution and higher total peak current through the protection element.

Bock and Hartnagel constructed two different field emission device ESD protection networks. In a first design, a two emitter structure was implemented with a 20 μm length, and a 250 nm tip radius. The emitter regions were coated with a gold film. In this design it was found that degradation was observed in the gold tip emitter structure. In this case, the FED ESD element did demonstrate an improvement in the RF device that was to be protected. This achieved ESD protection levels above 1600 V HBM in

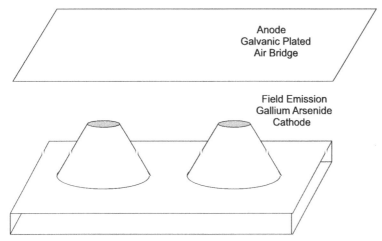

Anode
Galvanic Plated
Air Bridge

Field Emission
Gallium Arsenide
Cathode

Figure 6.4 Multi-emitter GaAs field emission device (FED)

GaAs prototypes (e.g., GaAs products typically do not achieve ESD HBM levels above 1000 V HBM) [7].

The second design of a FED ESD network design consisted of a ten emitter structure, with emitter lengths of 35 μm, and a 75 nm emitter tip radius. The emitter-to-emitter design pitch is 14 μm space. The trigger voltage was −7 V with the cathode negatively biased (e.g., emitter pointed wedge has a negative applied voltage relative to the anode air bridge with a positive bias), and 20 V when the cathode is positively biased (e.g., emitter pointed wedge has a positive applied voltage relative to the anode air bridge with a negative bias). It was pointed out that this implementation provided a bi-directional ESD network operation similar to an ESD "double-diode" network but with an asymmetric trigger condition [7].

With the utilization of the semiconductor process materials used in RF MMIC, the FED ESD element can be integrated into the physical design of a transistor element (Figure 6.4).

The above design can be improved by adding a control gate structure. Figure 6.5 shows a GaAs field emission triode. This structure uses a n+ GaAs grid to assist in the control of the field emission process, as well as dc currents.

A key ESD design practice in the implementation of Field Emission Device (FED) ESD networks is as follows:

- FED emitter-to-collector Gap control can lead to non-uniform conduction of emitters in multi-emitter design.

- FED current uniformity can be improved using a single semiconductor emitter instead of metal emitter.

- FED emitter ballasting can be improved by "electronic blunting effect" in multi-emitter FED structures.

- FED electronic blunting can compensate for emitter-to-collector gap control.

- FED devices can produce a bi-directional (but asymmetric trigger voltage) ESD solution.

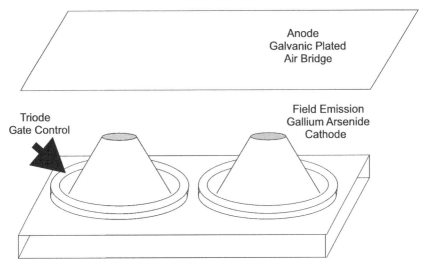

Figure 6.5   GaAs field emission device (FED) triode with control gate

In semiconductor chips, the upper surface of the semiconductor chip is passivated with films to avoid corrosion, contamination and scratching of films and interconnects. In RF semiconductors, and with the trend to micro-mechanical machines (MEMs), the usage of air bridges, and spark gaps is more acceptable than in the past. This will allow opportunities for implementation and integration of on-chip spark gaps and field emission devices (FED). Hence, because of the reliability, the repeatability, the time response, and the breakdown voltage, the usage of on-chip spark gaps and field emissions devices (FED) have a greater opportunity and potentially better success for implementation on-chip for high performance RF applications. Yet, process cost and other reliability issues may lead to future off-chip implementations.

6.8 SYSTEM LEVEL TRANSIENT SOLUTIONS

With increasing technology performance objectives, the ability to provide on-chip ESD protection will become more difficult. With the increase in circuit density, and smaller physical geometries, the ability to have on-chip ESD protection only, may be a limited perspective. Today, the trend toward finding both an on-chip and off-chip protection solution has increased. In systems today, there is an escalating number of signal lines and cable connections, where ESD events occur from the cards, boards and cables. With dynamically re-configurable systems, cards, boards and cables, system level solutions have to address environments which are too severe for an "on-chip ESD only" solution.

At the same time, the speed of advanced semiconductors may also limit the size of ESD protection networks. Both the allocated space and the allowed ESD capacitance budget will limit the circuit designers for on-chip ESD protection devices with the increased I/O count. The more stringent capacitance requirements create a potential performance-ESD reliability objectives conflict.

Off-chip voltage suppression devices will be a greater focus for cost, and perform-ance reasons. On a system level, circuit elements and devices used for ESD include off-chip low capacitance air gaps, resistors, capacitors, inductors, varistors, double-diode ESD elements and combinations thereof. Diodes were commonly used for over-shoot and under-shoot issues on cards, until the need for low capacitance, and cost were an issue. System-level diode capacitance in the order of 5 pF are used; these loading effects will be too high for future high frequency applications. Although diodes are capable of extremely low trigger voltages, the capacitance loading can also cause signal distortion. The need for economical ESD protection of multiple signal lines along with the need for low capacitance at high frequencies will be required for future high-pin count solutions.

Additionally, there are a number of present and future technologies where on-chip ESD protection is not feasible from a material perspective. For example, magnetic recording industry devices, such as magneto-resistor (MR) heads, giant magneto-resistor (GMR) heads, and tunneling magneto-resistor (TMR) heads are formed on non-silicon substrates.

An ESD design solution and practice is to incorporate multiple stages of ESD protection in system-level design. In these system-level solutions, the last level of ESD protection will be the on-chip ESD networks. These are further complimented by the addition of spark gaps, and surge protection concepts.

6.8.1 Transient Voltage Suppression (TVS) Devices

Off-chip Transient Voltage Suppression (TVS) devices are used to address system-level ESD events. Transient voltage suppressors (TVS) provide a low-cost solution to protect sensitive integrated circuits connected to high speed data and telecommunications lines from EOS, ESD and CDE (cable discharge events) that can enter through external plug ports and cause catastrophic damage. System level engineers are required to improve system-level perform-ance while maintaining the quality and reliability.

Electrostatic discharge (ESD), and electromagnetic emissions (EMI), are a concern in systems. System level standards and system engineers have long known that charged cables can also introduce system-level concerns. High voltage and high current events from charged cables are referred to as cable discharge events (CDE). Charge accumulation on untermi-nated twisted pair (UTP) cables occurs through both tribo-electric charging and induction charging. In the case of tribo-electrification, in a UTP cable can be dragged along a floor. A positive charge is established on the outside surface of the insulating film. The positive charge on the outside of the cable attracts negative charge in the twisted pair leads across the dielectric region. When the negative charge is induced near the outside positive charge, posi-tive charge is induced in the electrical conductor at the ends of the cable. As the cable is plugged into a connector, electrical arcing will occur, leading to a charging of the untermi-nated twisted pair (note: the twisted pair was neutral to this point). If a cable is introduced into a strong electric field, induction charging will occur. When the electric field is removed the cable remains charged until a discharge event from grounding occurs. Given a system which is un-powered, no latchup event can occur. But, if the cable discharge event occurs while a system is powered, the current event can lead to latchup. With the integration of

Wide Area Networks (WAN), and Local Area Network (LAN), the Ethernet is playing a larger role. When a charged twisted pair cable connects to an Ethernet port with a lower electrical potential, cable discharge events can occur in Local Area Network (LAN) systems. In the past, standards (e.g., IEEE 802.3 Section 14.7.2) have noted the potential for CDE processes in LAN cables. Additionally, the introduction of Category 5 and Category 6 cables has significantly low leakage across the dielectric. As a result, when a tribo-electric charge is established, the conductance of the insulator is so low that the induced charge can be maintained for long time scales (e.g., 24 hour period). In addition to the system level issues, the latchup robustness of advanced technology is significantly lower due to technology scaling of the latchup critical parameters. Hence, with both system and technology evolution, the reasons for the increased concern for this issue is the following:

- WAN and LAN integration.

- Category 5 and 6 LAN Cabling.

- Higher level incidents of disconnection.

- High level incidents of re-connections.

TVS devices can be used in many environments from one pair of high-speed differential signals (two lines) used in standard 10 BaseT (10 Mbits/sec), 100 BaseT (100 Mbits/sec), or 1000 BaseTX (1 Gbit/sec over copper) Ethernet transceiver interfaces. TVS devices sometimes use compensation diodes in a full-bridge configuration to reduce the capacitance loading effects, allowing signal integrity to be preserved at the high transmission speeds of GbE interfaces. TVS devices can be below 10 pF capacitance loading, and can achieve 100 A (8/20 μs) surge.

ESD design objectives of TVS devices include the following:

- Number of protection elements per signal pins.

- Low maximum reverse "standoff voltage."

- Low capacitance (1–10 pF).

- Low clamping voltages (near peripheral I/O power supply voltage of semiconductor chip levels).

- Achieve Telcordia GR-1089 intra-building lightning immunity requirements (100 A, 2/10-microsecond pulse durations).

- Peak pulse power maximum (in watts for 10–100 μs pulse duration time scales).

- Minimize board area.

Hence to address chip-level and system-level electrical overstress events (e.g., EOS, ESD, CDE, and lightning), transient voltage suppression (TVS) devices are used as a first line of protection. For the future, TVS devices will need to be low cost, low capacitance and be able to use a minimum of board area.

6.8.2 Polymer Voltage Suppression (PVS) Devices

The use of filled polymer for circuit protection was initiated in the early 1980s, with a self-resetting polymer fuse that had the capability of re-setting after a current surge [8,9]. In these structures, the materials increase in resistance as a function of current. Polymer positive temperature coefficient (PPTC) devices are used across all electronic markets because of their re-set feature and their formation into low profile films [8,9].

Polymer-based voltage suppression (PVS) devices can be used on a package, board, card or system level to suppress high current transient phenomenon [8,9]. To enable ESD protection of electronic products, polymer voltage suppression (PVS) devices are designed with low capacitance and space-efficient packages for multiple-line ESD protection. PVS devices incorporate system-level ESD protection without interfering with high frequency signals, and do not require space on printed circuit boards (PCB). The polymer voltage suppression devices can be built in array or matrix fashion to address multiple parallel pins in a system environment. [8]. The array of polymer voltage suppression devices can be incorporated in or in the physical system level connectors.

In polymer voltage suppression (PVS) components, the polymer provides a low dielectric constant. The dielectric constant of the polymer film can be formulated to meet a wide range of application frequencies. Figure 6.6 shows construction of a single polymer voltage suppression (PVS) surface-mount device consisting of a PVS film laminated between electrodes. Polymer voltage suppression devices are bi-directional, allowing the shunting of both positive and negative polarities from signal lines to ground locations. The polymer voltage suppression device film uses an epoxy polymer, containing uniformly-dispersed conductive and non-conductive particles, laminated between electrodes. The laminate is transformed into a surge protection device using a process similar to that used for manufacturing PCBs. The polymer voltage suppression (PVS) device overall thickness is in the range of less than 10 mils. The PVS material does not require a substrate and consequently, the trigger voltage of the device is set by device dimensions and the polymer formula. These components allow for single- or multiple-array surface mount components, and connector arrays. Installing the protection array on the connector itself provides ESD protection at the input traces of ESD sensitive components and semiconductors. K. Shrier demonstrated polymer voltage suppression elements with approximately 100 fF loading capacitance [8]. Polymer voltage

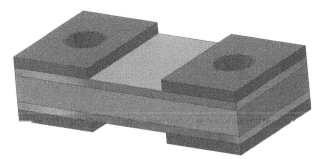

Figure 6.6 Polymer voltage suppression (PVS) device

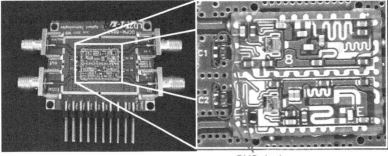
PVS devices

Figure 6.7   Polymer Voltage Suppression (PVS) devices integrated into cell phone application. Permission granted from ESD Association

suppression elements were achieved with less than 200 pF rise times, and effective clamping time of less than 10 ns [8,9]. Figure 6.7 shows the integration of the PVS device on a cell phone board with a GaAs cell phone application.

6.9 PACKAGE-LEVEL MECHANICAL ESD SOLUTIONS – MECHANICAL "CROWBARS"

In many ESD applications, ESD solutions are best addressed on the package, card, or board level instead of an on-chip ESD solution. In many applications, the design and application does not lend itself to integration with an ESD element. These can include semiconductor lasers, hetero-structures, super-lattices, Gallium Arsenide power amplifiers, and other low-pin count applications. In high-performance applications, the ability to provide on-chip ESD protection will become more difficult. With the increase in circuit density, and smaller physical geometries, the ability to have on-chip ESD protection only may be a limited perspective. ESD solutions on packages have been addressed since the 1960s with electrically connecting spring devices, such as proposed by Wallo [10]. Wallo [10], Kisor [11], Medesha [12], Tolnar and Winyard [13], Dinger *et al.* [14], Bachman and Dimeo [15], and Beecher [16] proposed mechanical shorting solutions to avoid electrostatic discharge in packages, static charge protective packages, connectors, cartridges and semiconductor packages. For example, in the 1990s, semiconductor laser diodes and ESD failure was common. A simple solution, developed by Cronin [17], to provide ESD protection between the three leads of a semiconductor diode was to form a helix-like mechanical spring that was inherently part of the packaging structure, forming an electrical short. When the laser element was not inserted, the mechanical spring shorted the leads. When the laser was inserted into the socket and the mechanical packaging spring was disengaged by mechanical deflection due to the insertion; this concept provided ESD protection when the element was removed from the application, and disengaged during application functional usage. This concept can be extended to other semiconductor packages [17–22], magnetic recording [23–28], and other applications that have an insertion process.

The fundamental off-chip ESD concept, independent of the application, is as follows:

- Mechanical shorting means can be applied between sensitive pins and ground references which are "shorted" when the application is not functional and "open" during functional usage.

- Mechanical shorting means can be applied between sensitive pins and ground references when the ESD sensitive element is not inserted into the card, board, or system.

- Mechanical shorting means can be initiated shorting all pins and all references when the application is not functional, and disengages when the application is in functional use.

- Mechanical "disengagement" of the shorting means is initiated by insertion, power on, or in the functional operation process.

6.10 DISK DRIVE ESD SOLUTIONS

6.10.1 In Line "ESD Shunt"

Many modern disk drives employ MR recording heads, also called "MR heads", "MR sensors", or "MR elements"; today it is well known that MR heads have a high susceptibility to damage from ESD. During operation of a magnetic storage drive, ESD is typically a relatively insignificant problem. The storage drive is usually encased within a computer, where it is protected from static discharge, particle contaminants, human interference, and other damage. In contrast, during the manufacture of magnetic storage drives, ESD can be a significant and perplexing problem, significantly reducing the effective yield of manufacturing operations. As a result, engineers are continually seeking effective ways to prevent ESD damage during manufacturing operations.

Traditionally, one of the best ways to reduce yield losses from ESD damage is to short the leads of an MR head together, creating an "electrical shunt." The electrical shunt used is the same film as the MR head itself. The MR head resistance is typically around 50 ohms; by providing a parallel 1 ohm shunt that is integrated with the MR head development avoidance of ESD failure can be achieved. This "MR shunt" provides an alternative electrical discharge path around the MR element, rather than through it. Experiments using a conventional MR head have shown that spanning the MR sensor with a 1 ohm connection increases the HBM failure voltage from 150 to 2000 volts. During the "lapping process" of the MR head, the MR ESD shunt is machined "off" to provide operability of the MR head for functional testing.

6.10.2 Armature – Mechanical "Shunt" – A Built-In Electrical "Crowbar"

. . . In 1996, I was standing in a line in Las Vegas at the EOS/ESD Symposium, waiting for lunch with A. Wallash and T. Hughbanks, when I proposed the idea to the two of them of integrating a mechanical "crowbar" across the leads of the MR head

in the armature structure of a disk drive – but did not understand how to activate and de-activate the closure of the crowbar switch. Not being an expert in MR heads, I missed a key point . . . Wallash and Hughbanks laughed, and said, "the armature rises when the MR head spins naturally." We found an disk drive armature expert, Satya Arya, and quickly launched this new invention. The patent attorneys told me that the US Patent Trademark Office (USPTO) called, and said they loved this idea; they stated that the best patents are "natural inventions" and simple.

Although the MR sensor is protected from ESD when its leads are shorted, this effectively renders the MR sensor inoperative. Therefore, to activate the MR sensor for manufacturing tests and the like, the shorted leads must be removed, disabled, or otherwise electrically disconnected. Likewise, after such tests, the interconnection between the leads must be reconnected to protect the MR sensor again. Manually shorting the leads in this manner, however, fails to provide a sufficiently convenient mechanism for protecting the MR head [26,27].

To protect the MR head in the disk drive itself, ESD protection can be built into the head disk assembly (HDA), and armature structure (Figure 6.8). ESD protection can be integrated into the disk drive assembly. By integrating an "ESD crowbar" a MR head receives ESD protection from a mechanism that automatically and releasably shorts the MR head whenever a suspension assembly on which the head is mounted is not installed in an HDA.

In the disk drive, the suspension assembly includes a flexure underlying a load beam, which is connected to an actuator arm. The MR head is mounted to a distal end of the flexure, leads from components of the MR head being brought out in the form of MR wire leads running along the load beam and the support arm to a nearby terminal connecting side tab. The conductors are separated and exposed at a designated point along the flexure to provide a contact region. A "shorting bar", which comprises an electrically conductive member attached to the actuator arm, automatically connects the MR wire leads at the contact region when absence of support for the MR head permits the load beam to bend sufficiently toward the shorting bar [26,27].

Thus, when the assembly is removed from installation in an HDA, the flexure is permitted to move toward the shorting bar, bringing the contact region and the shorting bar in electrical contact to short the MR wired leads and thereby disable the MR sensor. When the assembly is installed in an HDA, the MR head is supported by an air bearing or the disk itself,

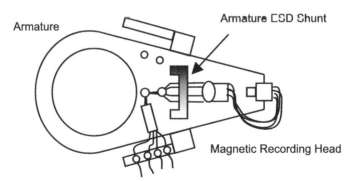

Figure 6.8 Armature with mechanical shunt device

depending upon whether the disk is rotating or stopped, respectively. In either case, the load beam is not permitted to droop and the shorting bar cannot contact the conductors, thus activating the MR sensor. Temporary ESD protection mechanisms are also provided, these being removable prior to operation of the HDA by breaking and removing various temporary shorting mechanisms [26,27].

6.11 SEMICONDUCTOR CHIP LEVEL SOLUTIONS – FLOOR PLANNING, LAYOUT, AND ARCHITECTURE

In semiconductor design, noise is a concern for both analog and radio frequency (RF) circuitry [33,49–53]. This is addressed using the following techniques:

- Spatial separation of different circuit domains.

- Guard ring structures between different circuit domains.

- Electrically isolated and separated power domains.

- Electrical connection between domains for ESD protection.

In the following sections, different architectures will be briefly discussed.

6.11.1 Mixed Signal Analog and Digital Floor Planning

In mixed signal design, a concern exists that the digital circuitry noise will impact the analog circuitry on a semiconductor chip level, and on a system level. In a mixed signal (MS) architecture, the digital and analog circuitry are separated into different power domains [52,53].

Figure 6.9 shows an example of a semiconductor chip with a digital and analog domain floorplan. To avoid ESD failures in a mixed signal (MS) semiconductor chip, ESD protection

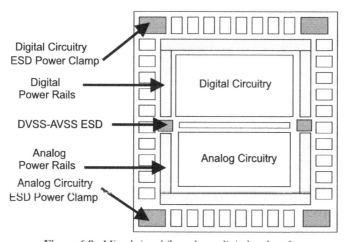

Figure 6.9 Mixed signal floorplan – digital and analog

networks are placed between the analog ground (AV_{SS}) and the digital ground (V_{SS}). Typical architectures contain a separate ESD power clamp in each domain. An ESD power clamp exists in the digital domain, between V_{DD} and V_{SS}, and a second ESD power clamp exists in the analog domain, between analog V_{DD} (AV_{DD}), and analog ground (AV_{SS}).

Alternative architectures are as follows:

- V_{DD}-to-AV_{DD} ESD Network: ESD network between the digital power rail (V_{DD}), and the analog power rail (AV_{DD}).

- V_{DD}-to-AV_{SS} ESD Network: ESD network between the digital power rail (V_{DD}), and the analog power rail (AV_{SS}).

The problem with this architecture is it allows for introduction of potential issues associated with EMI and EMC on the system level. Ground loops can be established given these different domains are only connected on a package level.

6.11.2 Bipolar-CMOS-DMOS (BCD) Floor Planning

Technologies that combine power transistors, bipolar and CMOS on a common substrate is known as Bipolar-CMOS-DMOS (BCD) technology [52,53]. DMOS transistors are typically high voltage transistors that are designed for power applications. These DMOS devices are designed to be operable for 5 V to 120 V operation. With the switching of large DMOS devices into the substrate, these can upset low voltage digital and analog circuitry. Hence it is necessary to floor plan semiconductor design with adequate separation between power transistors and sensitive low voltage circuitry. This is addressed using the following techniques:

- Spatial separation of the DMOS transistors from different circuit domains.

- Moats and guard rings around the DMOS devices themselves.

- Moats and guard ring structures between different circuit domains.

- Electrically isolated and separated power grids.

- Electrical connection between domains for ESD protection through the ground power rails.

6.11.3 System-on Chip Design Floor Planning

In a mixed signal (MS) architecture, the digital, analog and radio frequency (RF) circuitry are separated into different power domains [49–53]. Figure 6.10 shows an example of a semiconductor chip with a digital, analog, and RF domain. To avoid ESD failures in a mixed signal (MS) semiconductor chip, ESD protection networks are placed between the analog ground (AV_{SS}), the digital ground (V_{SS}), and the RF ground. Typical architectures contain a separate ESD power clamp in each domain. An ESD power clamp exists in the digital

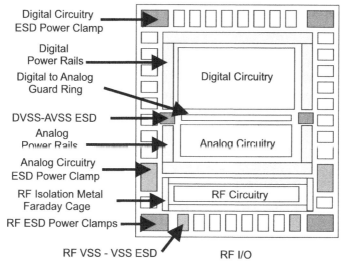

Digital Circuitry
ESD Power Clamp

Digital
Power Rails

Digital to Analog
Guard Ring

DVSS-AVSS ESD

Analog
Power Rails

Analog Circuitry
ESD Power Clamp

RF Isolation Metal
Faraday Cage

RF ESD Power Clamps

Digital Circuitry

Analog Circuitry

RF Circuitry

RF VSS - VSS ESD   RF I/O

Figure 6.10   Mixed signal architecture – digital, analog and RF architecture

domain, between V_{DD} and V_{SS}, and a second ESD power clamp exists in the analog domain, between analog VDD (AV_{DD}), and analog ground (AV_{SS}), and a third ESD power clamp is between RF VDD (RF V_{DD} or V_{CC}), and RF ground (RF V_{SS}, or V_{EE}). In these mixed signal chips, the RF application voltage is typically higher than the analog and digital application voltage.

Figure 6.10 shows an example of a mixed signal chip with RF, analog and digital circuitry [52]. To separate the analog circuitry from the digital noise, separate power rail domains exist. Additionally, a guard ring "moat" separates the two domains to produce a larger distance through the substrate region. The RF sector is separated on the lower sector of the chip floor plan. The RF circuitry is surrounded by layers of metal forming a "faraday cage" to isolate the RF signals. The faraday cage is formed by stacking the metal layers, and passing the signals through the breaks in the faraday cage. ESD network power clamps are placed in the digital, analog and RF domains between their power and ground rails. In addition, V_{SS}-to-V_{SS} ESD networks are placed to interconnect the ground rails. The V_{SS}-to-V_{SS} networks use series diode ESD elements, where the number of elements in the series is a function of the allowed capacitive coupling between the digital, analog, and RF sectors [52].

6.12 SEMICONDUCTOR CHIP SOLUTIONS – ELECTRICAL POWER GRID DESIGN

In this section, a chip level solution is discussed due to testing of a semiconductor chip with an IEC 61000-4-2 system pulse and the system-like HMM pulse. This discussion is how does one modify the internal design of a chip to avoid the propagation of the system pulse into the sensitive part of a semiconductor chip, or rephrased, how does one design semiconductor chips so that the system tests do not bring down the components?

6.12.1 HMM and IEC Specification Power Grid and Interconnect Design Considerations

For human metal model (HMM) and IEC 61000-4-2 specifications, the peak current can exceed 30 to 40 A [41–45]. In the HMM and IEC 61000-4-2 specifications, only pins which are connected to external ports are required to receive this high current pulse. Secondly, the large current in the substrate can influence the non-IEC tested circuitry. To protect the IEC tested pins, and avoid failure of the non-IEC pins, the IEC pins can be isolated in the power grid [52]. This can be achieved by the following means:

• Independent IEC power domain: (IEC signal pin, IEC V_{DD} bus, and IEC V_{SS} bus).

• Dual width power bus: (IEC and Non-IEC domains).

• Resistance Segmentation: Resistance separated IEC and Non-IEC domains.

Figures 6.11 and 6.12 shows examples of separate IEC vs non-IEC power grids, and connectivity [52].

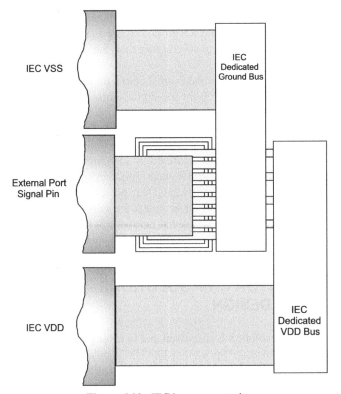

Figure 6.11 IEC bus segmentation

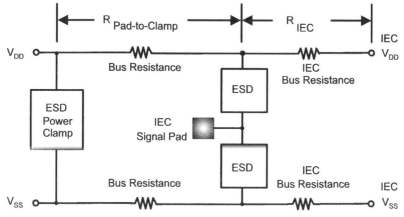

Figure 6.12   IEC Bus power and ground placement

6.12.2 ESD Power Clamp Design Synthesis – IEC 61000-4-2 Responsive ESD Power Clamps

For applications that are required to respond to the IEC 61000-4-2 pulse event, not all circuit topologies are suitable [52]. For the IEC 61000-4-2 event, there is a fast current pulse which is of considerable magnitude. Hence, to address the frequency response and current magnitude, many ESD power clamps are required to be modified.

During the IEC 61000-4-2 event on the chassis or ground line of a system, a negative pulse occurs on the V_{SS} power rail or substrate. This can initiate the RC-triggered network from the negative pulse event. But, the elements in the RC discriminator must be responsive, or circuit failure can occur. The resistor and capacitor element choices must be responsive. Resistors, such as polysilicon resistors, may be slow to respond to fast events.

Figure 6.13 is an example of an IEC 61000-4-2 event responsive ESD MOSFET network. The advantage of this network is that the p-channel MOSFET is more responsive than a

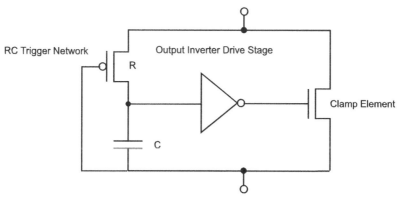

Figure 6.13   IEC 61000-4-2 responsive ESD power clamp

polysilicon resistor element. Additionally, so that the inverter drive network is more responsive, only a single inverter stage is implemented [52].

6.13 ESD AND EMC – WHEN CHIPS BRING DOWN SYSTEMS

ESD component and system level testing are typically not done together. Component level ESD testing is done on components, and system level ESD testing was done on a system level. As discussed in the prior chapters, a first set of ESD tests are completed for the components (e.g., HBM, MM, and CDM), and a different set are being applied to systems (e.g., IEC 61000-4-2, CDE, CBM).

In the next section, the following issues will be addressed:

• How does device level ESD testing degradation influence functional system-level tests?

• How does ESD component level robustness relate to system level susceptibility?

• How do you bridge from a semiconductor ESD robustness to a system susceptibility when today they are not practicing the same test?

To answer the first question, a method will be shown on how to bridge between device level degradation and system level response.

As for the next questions, a new method is shown that can be carried out by the device and system manufacturer and correlated to system level upsets, and therefore determine the location and cause of the upset.

6.14 SYSTEM LEVEL AND COMPONENT LEVEL ESD TESTING AND SYSTEM LEVEL RESPONSE

In this first methodology, an impedance method is shown that correlates impedance shifts, or degradation using time domain reflection (TDR) methods; this is followed by insertion of the "degraded component" into the system to evaluate how this degradation impacts the system metrics and measurements.

6.14.1 Time Domain Reflection (TDR) and Impedance Methodology for ESD Testing

Signal integrity is important to both digital systems and RF communication systems as clock frequencies and data transmission rates increase. For RF and high data rate transmission systems, new testing techniques are needed to evaluate the impact of ESD on components. At these high data transmission rates, ESD-induced changes to the electronics that impact signal integrity can lead to impacts on the system's reliability. ESD damage, either permanent or latent, can lead to chip or performance issues. ESD induced damage influences the gain, the transconductance, and the S-parameters, a new methodology and failure criteria may be needed to evaluate the ESD impacts. This can occur to circuitry on the transmission or the receiving end of a system leading to unacceptable degradation levels. ESD-induced damage can include the following circuit and system issues:

- Signal rise time.

- Pulse width.

- Timing.

- Jitter.

- Signal-to-noise ratio.

In high speed communication systems and high speed components, the transmission characteristics are important for the transference of signals. The transmission and reflection coefficients are key measures of the operation and functionality of the system. The transmission coefficients, the S-parameters and the impedance play a large role in the functionality. Shifts in the S-parameters or transmission coefficients can be quantified using the Time Domain Reflectometry (TDR) methodology.

The Time Domain Reflectometry (TDR) method can be used to verify ESD failures and implications to the functional system. The TDR method is common practice in high speed system development. The functional test requires a launching of a signal and the reflected wave is measured to evaluate the transmission, the reflection and impedance of the port tested. Time domain reflectometry measures the reflections that occur from a signal, which is traveling through a transmission environment. The transmission environment can be a semiconductor circuit, a connector, cable or circuit board. The TDR instrumentation launches a signal or pulse through the system to be evaluated and compares the reflections of a standard impedance to that of the unquantified transmission environment.

Figure 6.14 shows an example of a TDR measurement system. A TDR measurement sampling module consists of a step source, a 50 Ω connection, followed by a transmission line to the load. The TDR sampling module also contains a sampler circuit, which draws a signal off a 50 Ω transmission line, whose signal is fed back to the oscilloscope. The TDR display is the voltage waveform that is reflected when a fast voltage step signal is launched down the

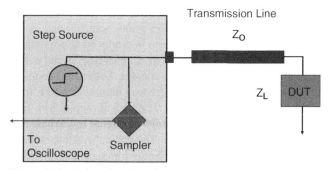

Figure 6.14   A time domain reflectometry (TDR) test methodology

transmission line. The waveform, which is received at the oscilloscope, is the incident step as well as the set of reflections generated from impedance mis-match and discontinuities in the transmission system.

The mathematics of the TDR method is based on the impedance ratios and a reflection coefficient ρ. The reflection coefficient ρ is equal to the ratio of the reflected pulse amplitude to the incident pulse amplitude

$$\rho = \frac{V^*_{reflected}}{V_{incident}}$$

The reflection coefficient can be expressed as a function of the transmission line characteristic impedance Z_O and the fixed termination impedance Z_L. In this form, the reflection coefficient can be expressed as

$$\rho = \frac{Z_L - Z_O}{Z_L + Z_O}$$

The fast-step stimulus waveform is delivered to the Device Under Test (DUT) after propagation through the sample head, the transmission line, connectors and the test fixture connections. The waveform which is reflected from the device under test is delayed by the two electrical lengths at the oscilloscope – the time of flight through all the interconnects and the return flight time. This signal is superimposed on the incident waveform at the TDR sampling head. The TDR sampling heads typically allow evaluation of the voltage waveform, the reflection coefficient or the impedance on the TDR oscilloscope.

6.14.2 Time Domain Reflectometry (TDR) ESD Test System Evaluation

Time domain reflectometry (TDR) is valuable for the analysis of ESD induced failure in both components or systems. TDR methodology can be valuable when only the input is available for analysis. As an example of application of the TDR system and ESD-induced degradation, a high speed optical interconnect system is shown [49,50,55].

Figure 6.15 shows a high level diagram of the optical interconnect system. The system consisted of a transmitter/receiver module, a short-wave Vertical Cavity Surface Emitting Laser (VCSEL), an optical wave guide, an optical photo-detector, and an Optical-to-Electrical (O/E) converter. The data input signal stream is produced at approximately 2 Gb/s in a Fiber Channel Pattern (FCPAT). The data pattern is generated by the pattern generator and converted to a differential signal via the hybrid coupler. The hybrid coupler element is electrically connected to the transceiver (TX) differential inputs. The SFF transceiver receives the data pattern and modulates the laser diode. The optical output from the SFF transceiver is connected to a multi-mode fiber, which conveys the signal to an optical attenuator before making its way to the optical input of the DCA. The Optical-to-Electrical (O/E) converter in the DCA filters the waveform by limiting the bandwidth and projects the displayed waveform on screen.

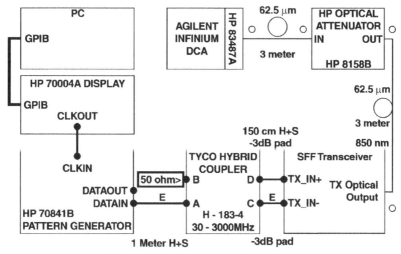

Figure 6.15   Electro-optical test system

Optical-to-electrical conversion in the transceiver is a very precise function requiring tight control of parameters for gigabit data transfer. The electrical path from the connector to the MICC chip is a controlled 50 Ω single-ended impedance. A transceiver having this form factor has exposed transmitter input pins exposed to ESD events. ESD events can damage the circuits and destroy the circuit integrity. Testing the O/E conversion before and after ESD events is a way of monitoring the robustness of the ESD protection circuits.

The ESD test methodology for evaluation of the system comprised of the following steps [49,50,55]:

• Functional characterization of the system is performed prior to ESD-stress.

• The input TDR signal is evaluated in an un-powered state.

• System power is applied.

• Optical "eye" patterns are recorded prior to ESD testing.

• ESD stress is applied to the transceiver chip input signals.

• An "ESD gun" or ESD pulse system is directly connected to the subject pin and the stimulus is applied.

• Each pin on the evaluation card is connected to signals, power, and grounds in the transceiver.

• Differential input TX_IN pins would be source signal and the return signal path was either the power or ground pin. Both positive and negative polarities are tested.

• TDR measurements were evaluated post-ESD stress. An ESD impulse is applied to the port when the system is un-powered; the system is then re-tested using the TDR test methodology.

• Post-ESD stress output "eye" test is evaluated to observe system level degradation effects.

Prior to ESD testing, the functional system is characterized. The experiment is started by observing the input TDR signal when the system is powered down. The TDR system consists of a Tektronix SD24 TDR Sampling head in a Tektronix 11801C Oscilloscope. The Tektronix 20 GHz SD24 TDR Sampling plug-in has a 15 ps rise time into a load and a 35 ps reflected wave rise time. The Tektronix 11801C has a 50 GHz sampling rate. After power-up the data input signal stream is initiated, and the output optical eye patterns are recorded prior to any ESD testing. The transceiver/receiver chip was first analyzed followed by full system evaluation.

The ESD test method establishes a procedure to apply ESD pulses to all pads on the transceiver via externally placed pins. An "ESD gun" Mini-Zap 2000 is directly connected to the subject pin and the stimulus is applied in accordance to the JEDEC Standard (JESD22-A114-B). The ESD gun Mini-Zap 2000 is wired to the desired pins for source and return. Each pin on the evaluation card is connected to signals, power, and grounds in the transceiver. In this particular test, one of the TX_IN pins would be source signal and the return signal path was either the power or ground pin. Note that both positive and negative ESD pulse polarities can be evaluated. The source/return path from the "ESD gun" Mini-Zap 2000 lead is in the following order: stimulus pin, card trace, card connector, transceiver connector, transceiver trace/component, and finally MICC chip input (TX_IN). An ESD pulse stress is applied to the port when the system is un-powered; the system is then re-tested using the TDR test methodology. An example of the results below shows the characteristics of a system of a 1 GHz path for stress at 2000 V and above 2000 V.

In the test system, the semiconductor chip which may be vulnerable to ESD events is the SFF transceiver chip which consists of radio frequency (RF) components. In this application, a 45 GHz f_T Silicon Germanium (SiGe) hetero-junction bipolar transistor technology is used in the transceiver chip. Figure 6.16 shows a high-level diagram of the transceiver chip architecture. In the diagram, the differential inputs, the amplifiers, and the diode laser signals are shown.

Measurements were taken at different stress levels, evaluating the TX-pin distortion. HBM stressing was performed on the transceiver chip to determine the ESD sensitivity of the signal pins. Table 6.1 shows a table of transmission and reflection signal magnitudes at a magnitude of 2000 V HBM ESD stress. Each line in the table represents a different module placed under ESD stress. Various system level parameters of amplitude and loss are recorded

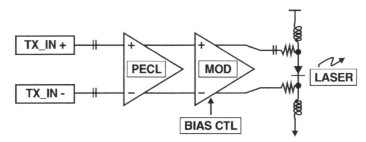

Figure 6.16 Transceiver chip architecture

Table 6.1 TDR Method Results at 2000 V HBM ESD Stress

Type	Serial	FCP	TX ER	TX DJ	1e-12 RX Sensitivity	RX Amplitude	TX Fault	Rx Loss
1 GHz fc path	1505	−5.69	7.40	34.78	−18.7	756.06	ok	ok
	2735	−5.64	6.62	22.17	−19.2	761.03	ok	ok
	2779	−6.12	7.25	24.01	−19.7	743.39	ok	ok
	2750	−6.48	5.96	18.3	−18.7	717.93	ok	ok
	1502	−5.82	8.82	45.45	−19.8	740.10	ok	ok

for the transmitted and reflected signals. Below a 2000 V HBM level, distortions in the TX amplitude is not observed in the 1 GHz signal path. In Table 6.2, the ESD stress was increased above 2000 V HBM. From the TDR method, various parameters begin to distort as the ESD stress increases. Above 2000 V, shifts occur in the transmission port tested. Distortion of the transmitter is noted in the TX pin with a 2X increase in TX DJ (TX DJ increased from levels of 30 to 72 and 131). Hence, from this TDR methodology, the metrics are evaluated for the various signal level parameters [49,50].

A second means of observation is the output "eye" test. After the input transmitter/receiver circuit has its TDR measurement, the system is re-powered and cycled at the 2 Gb/s data rate. From the output "eye" the distortion of the output characteristics is another way of evaluation of the ESD impacts. The timing of the optical network is impacted by the ESD pulse as a result of the impact on the differential input of the input SFF chip. Experimental observations showed that, in some cases, it was hard to determine the change in the TDR input characteristic before and after ESD stress yet observations were visible in the subtle changes in the output "eye test" [49,50,55].

Hence, using a Time Domain Reflectometry (TDR) technique, and system level "eye test," the impact on a RF chip at the input and output of the system can be evaluated. The TDR method, using state-of-the-art oscilloscopes with TDR sampling heads, can evaluate

Table 6.2 TDR Method Results above 2000 V HBM ESD Stress

Type	Serial	FCP	TX ER	TX DJ	1e-12 RX Sensitivity	RX Amplitude	TX Fault	Rx Loss	Comments
1 GHz path	1505	−5.69	7.97	131.48	−18.8	770.85	ok	ok	TX-distorted
	2735	−5.64	6.98	24.58	−19.4	771.89	ok	ok	TX – ok
	2779	−6.12	6.75	28.85	−19.6	767.84	ok	ok	TX – ok
	2750	−6.48	7.25	30.57	−18.8	786.23	ok	ok	TX – ok
	1502	−5.82	7.68	72.53	−19.8	787.76	ok	ok	TX – hi crossing

the reflection coefficients, the impedance and voltage level response. Using the methodology of TDR measurements of the system input (in an un-powered state) pre- and post-ESD allows for a comparison evaluation of the stress on the TDR waveform. It was also found that the "eye test" can observe low voltage degradation effects not solely observed from the TDR method. With a trained eye, the subtle variations in the "eye test" can allow a test engineer the means of observing distortions which may impact the data transmission system.

6.14.3 ESD Degradation System Level Method – Eye Tests

In system level quantification of ESD degradation, it is not always possible to determine the ESD-induced degradation at the semiconductor chip, or at low-speed characterization. Using the "eye test", it is possible to provide quantification of the system level impact of ESD degradation. A means of observation is the output "eye" test. From the comparison of the output "eye" before and after ESD degradation, system level failure criteria can be established. From the output "eye" the distortion of the output characteristics is another way of evaluation of the ESD impacts. The eye test is a measure of the timing of signals to evaluate worst case operational means. Figure 6.17 is a pictorial example of the "eye" before and after ESD stress. The "eye" is formed by the overlaying of two signals of interest where one signal is inverted. When the timing is sound, the "eye" is open and wide. During distortion or poor timing, the overlapping of the signal leads to a small eye opening. Below is an example of an "eye test" whose iris has been degraded as a result of ESD induced degradation.

Figure 6.18 shows an example of an "eye test" with a system with good functional characteristics. Figure 6.19 shows the results of the output "eye" after ESD testing.

Experimental observations will show that in some chip applications it will be difficult to determine the change in the semiconductor chip input characteristics, before and after ESD

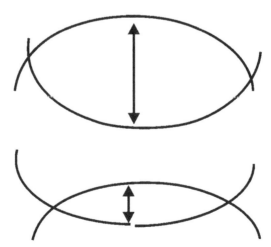

Figure 6.17 Functional eye test before and after ESD testing. The first characteristic represents a "good eye" and the second represents a "bad eye"

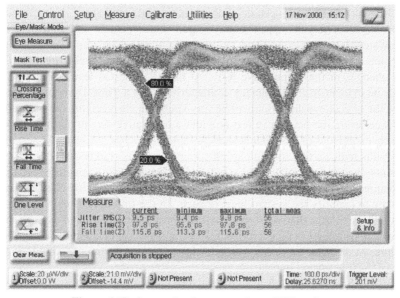

Figure 6.18   System level eye test prior to ESD testing

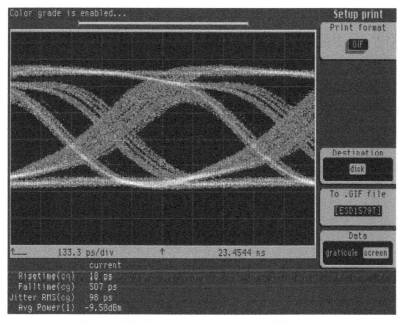

Figure 6.19   System level eye test post-ESD testing

stress. At low current stress levels, dc characteristic shifts, and small degradations in the device characteristics may not appear to have significant impact on the semiconductor chip. But, on a system level these can lead to system failure. In some applications, the semiconductor chip level ESD degradations may not be apparent, but observations are visible in the subtle changes in the output "eye test". At high ESD stress levels where significant change can be evident in the dc device characteristics, or a Time Domain Reflectometry (TDR) test, the optical eye distortion will be very visible in the receiving output oscilloscope signals. Hence, using the output eye test provides a qualitative way to observe distortion of the device characteristics of the input devices [49,50,55].

These results indicate that for high-speed communications systems, new criteria and test techniques may be needed to define the impact of ESD stress. Traditional ESD testing looks at the dc parameter shifts of the semiconductor device or input pin chip characteristics may be inadequate to quantify the system level impacts. In the future, ESD degradation may be best quantified using quantitative time-domain reflection techniques and evaluations of the impedance shifts, reflection characteristics and transmission characteristics to determine device failure. At the same time, this can be further understood using the output "eye" test in evaluation of the true system impact.

From this method, component level degradation can be correlated to system level susceptibility and system level performance impact. A correlation can be developed between the impedance degradation, and how it influences the system level performance objectives of rise time, fall time, jitter, and eye opening. This methodology provides the component-to-system level correlation, but does not provide spatial or visual means. In the next methodology, a means of visualizing the location of the performance degradation can be evaluated.

6.15 EMC AND ESD SCANNING

A methodology is needed for both component and system level manufacturers to test for susceptibility of both the components and the systems, and have a means to correlate the relationship. A susceptibility test that can correlate system level upsets, and the location of the upset is key to building better systems.

Today, component ESD models, which are unpowered, can not provide enough information to estimate the performance at a "powered" state of the system.

In a printed circuit board (PCB), mother boards, and the components themselves, the ability to determine where the susceptibility problem occurs is very important. With the complexity of computer chips, and the spatial extent of the system board, it is very important to determine exactly where the problems occur in order to improve and fix them. One of the biggest problems today is that there is no visualization capability of the location of the susceptibility concern. As a result, electromagnetic compatibility (EMC) and susceptibility improvements are evaluated by relying on both trial-and-error experimentation, and EMC experience.

Electromagnetic compatibility (EMC)/electrostatic discharge (ESD) scanning is a method developed to determine the susceptibility or upsets of systems, circuits, and components to ESD or other EMC events without causing hard failures [56–59]. In order to provide visualization of the EMC/ESD sensitivity, a localized source and means of scanning over the system is important to evaluate the local system or circuit response.

An ESD/EMC scanning system has been developed that produces a localized E-field or H-field, scans the equipment under test (EUT), and monitors the product for disturbances or upsets. This test can scan the exterior of products, sub-assemblies, and boards. The same test can be applied to components within the system (e.g., semiconductor integrated circuits). In both cases, a mapping can be produced which overlays the system to determine entry points of electromagnetic noise, and system sensitivity.

This ESD/EMC scanning system can apply different stimulus sources [56–59]. The pulse test can be an RF source, electrical fast transients (EFT), transmission line, or ESD gun. A transmission line pulse (TLP) source can be used to provide a transmission line pulse (TLP) or very fast transmission line pulse (VF-TLP) event. For simulation of system-like events, the human metal model (HMM) or IEC 61000-4-2 test can be applied.

Figure 6.20 shows an example of an ESD/EMC scanning system [56–59]. Figure 6.21 shows a high level diagram of the major sub-systems in the ESD/EMC scanning system. Figure 6.21 shows the sample table area, the position arms, the motor drivers, data collection system, and controlling computer. The high level diagram also shows the high voltage supply, a TLP source, and the connection to the position arms. To produce a local H-field, the tip of the probe is formed into a small loop. The pulse from the TLP system initiates a pulse current, which generates the local H-field.

After the scanning process, the system produces a mapping of the susceptibility overlaid on the physical image of the board, or components. Figure 6.22 shows a product sensitivity mapping of two products on the same system. It can be observed that the image mapping on the left (Vendor A) is more sensitive compared to the image mapping

Figure 6.20 EMC/ESD scanning system. Permission granted from Amber Precision Instruments, Inc.

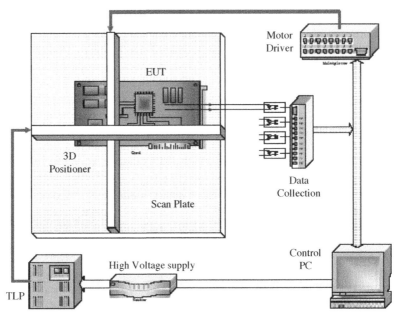

Figure 6.21 EMC/ESD scanning system high level diagram. Permission granted from Amber Precision Instruments, Inc.

on the right (Vendor B). Although both fulfill the same electrical functional specification, their sensitivity to EMI is not equivalent.

Figure 6.23 is an example of a motherboard. In the image mapping, there are regions of sensitivity. With this mapping, it provides guidance for potential re-design of the motherboard [56–59].

With this EMC/ESD scan system, evaluation of components and system boards can be evaluated in the design phase, assembly, or for qualification. A convergence of system-level and chip level ESD testing is occurring today with these new methodologies.

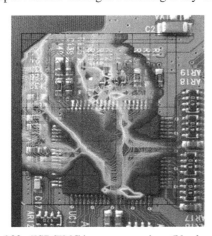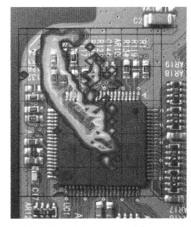

Figure 6.22 ESD/EMC image comparison (Vendor A and B). Permission granted from Amber Precision Instruments, Inc.

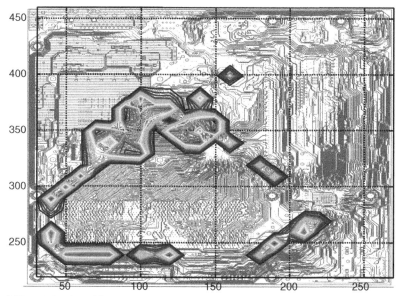

Figure 6.23   Mother board image scan. Permission granted from Amber Precision Instruments, Inc.

6.16 SUMMARY AND CLOSING COMMENTS

Chapter 6 addressed the issue of non-traditional ESD networks and on-chip and off-chip protection networks. Spark gaps, field emission devices and conductive polymer ESD protection concepts have been demonstrated in Gallium Arsenide technology and applications with significant success. The transition from non-traditional semiconductor ESD networks, and off-chip ESD protection may be required as the application frequencies continue to increase, and chip sizes decrease. It is unclear at this time which technology type and technology nodes will require these solutions, and when these become a standard solution for all semi-conductor products. Additionally, in order for these solutions to continue, the scaling of these elements (e.g., in trigger voltages, on-currents, on-resistance, size, and cost) is crucial to further improve them for future generations.

New testing methodologies are discussed to evaluate system level response to component level ESD degradation, as well as sensitivity to electromagnetic fields. Two examples are shown, where the "degraded component" was placed into the system to evaluate the system level performance degradation. Additionally, an ESD/EMC scanning system was shown that addresses both component and system level susceptibility. This new methodology bridges the gap between component level ESD evaluation and EMC system level testing. Also, this ESD/EMC test method provides a mapping of the locations of susceptibility, finding the root cause, and allowing for re design or improvements.

Chapter 7 will address ESD issues in future nanosystems. The chapter discusses ESD problems in photo-masks, magnetic recording devices, micro-motors, micro-mirrors, RF MEM switches, FinFETs, to nanowires.

REFERENCES

1. Paschen, F. (1889) Ueber die zum Funkenübergang in Luft, Wasserstoff und Kohlensäure bei verschiedenen Drucken erforderliche Potentialdifferenz, *Annalen der Physik*, **37**, 69.
2. Toepler, M. (1906) Über Funkenspannungen, *Annalen der Physik*, **4**, 191.
3. Von Hippel, A. (1965) Conduction and breakdown, in *The Molecular Designing of Materials and Devices*, MIT Press, Boston, 183–197.
4. Townsend, J.S. (1915) *Electricity in Gases*, Clarendon Press, Oxford.
5. Kleen, B.G. (1972) Printed circuit spark-gap protector. *IBM Technical Disclosure Bulletin*, **4** (2), 638.
6. DeBar, D.E. *et al.* (1975) Module spark gap. *IBM Technical Disclosure Bulletin*, **18** (7).
7. Bock, K. (1997) ESD issues in compound semiconductor high-frequency devices and circuits. Proceedings of the Electrical Overstress/Electrostatic Discharge (EOS/ESD)Symposium, pp. 1–12.
8. Shrier, K., Truong, T., and Felps, J. (2004) Transmission line pulse test methods, test techniques, and characterization of low capacitance voltage suppression device for system level electrostatic discharge compliance. Proceedings of the Electrical Overstress/Electrostatic Discharge (EOS/ESD) Symposium, pp. 88–97.
9. Shrier, K. and Jiaa, C. (2005) ESD enhancement of power amplifier with polymer voltage suppressor. Proceedings of the Taiwan Electrostatic Discharge Conference, pp. 110–115.
10. Wallo, W.H. (September 16 1967) Electrically connecting spring device. U.S. Patent No. 3,467,930.
11. Kisor, T.W. (April 4 1972) Static charge protective packages for electronic devices. U.S. Patent No. 3,653,498.
12. Medesha, A.L. (November 20 1973) Package including electrical equipment lead shorting element. U.S. Patent No. 3,774,075.
13. Tolnar Jr., E.J. and Winyard, A.H. (March 4. 1975) Connector means having shorting clip. U.S. Patent No. 3,869,191.
14. Dinger, E.D., Saben, D.G., and VanPatten, J.R. (April 19 1977) Static control shorting clip for semiconductor package. U.S. Patent No. 4,019,094.
15. Bachman, W.J. and Dimeo, F.R. (December 18 1979) Plug-in circuit cartridge with electrostatic charge protection. U. S. Patent No. 4,179,178.
16. Beecher, R. (July 23 1985) Cartridge having improved electrostatic discharge protection. U.S. Patent No. 4,531,176.
17. Cronin, D.V. (November 20 1990) Electrical connector with attachment for automatically shorting select conductors upon disconnection of connector. U.S. Patent No. 4,971,568.
18. Voldman, S. (2002) Lightning rods for nanoelectronics. *Scientific American*, **287** (4), 90–97.
19. Cronin, D.V. (April 28 1992) Electrostatic discharge protection devices for semiconductor chip packages. U.S. Patent No. 5,108,299.
20. Cronin, D.V. (November 17 1992) Electrostatic discharge protection devices for semiconductor chip packages. U.S. Patent No. 5,163,850.
21. Cronin, D.V. (November 17 1992) Electrostatic discharge protection device for a printed circuit board, U.S. Patent No. 5,164,880.
22. Johansen, A.W. and Cronin, D.V. (September 22 1998) Electrostatic discharge protection device. U.S. Patent No. 5,812,357.
23. Wallash, A., Hughbanks, T., and Voldman, S. (1995) ESD failure mechanisms of inductive and magnetoresistive recording heads. Proceedings of the Electrical Overstress/Electrostatic Discharge (EOS/ESD) Symposium, pp. 322–330.
24. Bajorek, C.H., Erpelding, A.D., Garfunkel, G.A. *et al.* (November 7 1995) Shorted magnetoresistive head leads for electrical overstress and electrostatic discharge protection during manufacture of a magnetic storage system. U. S. Patent No. 5,465,186.

25. Hughbanks, T.H., Lee, H.P., Phipps, P.B. *et al.* (February 13, 1996) Shorted magnetoresistive head elements for electrical overstress and electrostatic discharge protection. U.S. Patent No. 5,491,605.
26. Arya, S.P., Hughbanks, T.S., Voldman, S.H., and Wallash, A.J. (July 1, 1997) Electrostatic discharge protection system for MR heads. U.S. Patent No. 5,644,454.
27. Arya, S.P., Hughbanks, T.S., Voldman, S.H., and Wallash, A.J. (January 20, 1998) Electrostatic discharge protection system for MR heads. U.S. Patent No. 5,710,682.
28. Jowett, C.E. (1976) *Electrostatics in the Electronic Environment*, Halsted Press, New York.
29. Lewis, W.H. (1995) *Handbook on Electromagnetic Compatibility*, Academic Press, New York.
30. Morrison, R. and Lewis, W.H. (1990) *Grounding and Shielding in Facilities*, John Wiley and Sons Inc., New York.
31. Paul, C.R. (2006) *Introduction to Electromagnetic Compatibility*, John Wiley and Sons Inc., New York.
32. Morrison, R. and Lewis, W.H. (2007) *Grounding and Shielding*, John Wiley and Sons Inc., New York.
33. Ott, H.W. (2009) *Electromagnetic Compatibility Engineering*, John Wiley and Sons Inc., Hoboken, New Jersey.
34. Ott, H.W. (1985) Controlling EMI by proper printed wiring board layout. Sixth Symposium on EMC, Zurich, Switzerland.
35. Radio Technical Commission for Aeronautics (RTCA) RTCA/DO-160E (December 7 2004) *Environmental Conditions and Test Procedures for Airborne Equipment*, Radio Technical Commission for Aeronautics (RTCA).
36. Society of Automotive Engineers SAE J551 (June 1996) *Performance Levels and Methods of Measurement of Electromagnetic Compatibility of Vehicles and Devices (60 Hz to 18 GHz)*, Society of Automotive Engineers.
37. Society of Automotive EngineersSAE J1113 (June 1995) *Electromagnetic Compatibility Measurement Procedure for Vehicle Component (Except Aircraft) (60 Hz to 18 GHz)*, Society of Automotive Engineers.
38. Wall, A. (2004) Historical perspective of the FCC rules for digital devices and a look to the future. IEEE International Symposium on Electromagnetic Compatibility, August 9–13, 2004.
39. Denny, H.W. (1983) *Grounding For the Control of EMI*, Don White Consultants, Gainesville, VA.
40. Boxleitner, W. (1989) *Electrostatic Discharge and Electronic Equipment*, IEEE Press, New York.
41. International Electro-technical Commission (IEC) IEC 61000-4-2 (2001) Electromagnetic Compatibility (EMC): Testing and Measurement Techniques – Electrostatic Discharge Immunity Test.
42. Grund, E., Muhonen, K., and Peachey, N. (2008) Delivering IEC 61000-4-2 current pulses through transmission lines at 100 and 330 ohm system impedances. Proceedings of the Electrical Overstress/Electrostatic Discharge (EOS/ESD) Symposium, pp. 132–141.
43. IEC 61000-4-2 (2008) Electromagnetic Compatibility (EMC) – Part 4-2:Testing and Measurement Techniques – Electrostatic Discharge Immunity Test.
44. International Electro-technical Commission (IEC)IEC 61000-4-2 (2001) Electromagnetic Compatibility (EMC): Testing and Measurement Techniques – Electrostatic Discharge Immunity Test.
45. Grund, E., Muhonen, K., and Peachey, N. (2008) Delivering IEC 61000-4-2 current pulses through transmission lines at 100 and 330 ohm system impedances. Proceedings of the Electrical Overstress/Electrostatic Discharge (EOS/ESD) Symposium, pp. 132–141.
46. IEC 61000-4-2 (2008) Electromagnetic Compatibility (EMC) – Part 4-2:Testing and Measurement Techniques – Electrostatic Discharge Immunity Test.
47. Voldman, S. (2004) *ESD: Physics and Devices*, John Wiley and Sons, Ltd., Chichester, England.
48. Voldman, S. (2005) *ESD: Circuits and Devices*, John Wiley and Sons, Ltd., Chichester, England.
49. Voldman, S. (2008) *ESD: Circuits and Devices*, Publishing House of Electronic Industry (PHEI), Beijing, China.

50. Voldman, S. (2006) *ESD: RF Circuits and Technology*, John Wiley and Sons, Ltd., Chichester, England.
51. Voldman, S. (2011) *ESD:RF Circuits and Technology*, Publishing House of Electronic Industry (PHEI), Beijing, China.
52. Voldman, S. (2009) *ESD: Failure Mechanisms and Models*, John Wiley and Sons, Ltd., Chichester, England.
53. Voldman, S. (2011) *ESD: Design and Synthesis*, John Wiley and Sons, Ltd., Chichester, England.
54. Vashchenko, V. and Shibkov, A. (2010) *ESD Design in Analog Circuits*, Springer, New York.
55. Voldman, S. (2007) *Latchup*, John Wiley and Sons, Ltd., Chichester, England.
56. Voldman, S., Ronan, B., Ames, S. *et al.* (2002) Test methods, test techniques and failure criteria for evaluation of ESD degradation of analog and radio frequency (RF) Technology. Proceedings of the Electrical Overstress/Electrostatic Discharge (EOS/ESD) Symposium, October 2002, pp. 92–100.
57. Pommerenke, D., Koo, J., and Muchaidze, G. (Feb. 2006) Finding the root cause of an ESD upset event. *DesignCom 2006*, Santa Clara.
58. Pommerenke, D., Muchaidze, G., Min, J. *et al.* (2007) Application and limits of IC and PCB scanning methods for immunity analysis. Proceedings of the 18th Int. Zurich Symposium on Electromagnetic Compatibility (EMC), Munich.
59. Muchaidze, G., Koo, J., Cai, Q. *et al.* (2008) Susceptibility scanning as a failure analysis tool for system-level electrostatic discharge problems. *IEEE Transactions on Electromagnetic Compatibility*, **50** (2), 268–276.

7 Electrostatic Discharge (ESD) in the Future

7.1 WHAT IS IN THE FUTURE FOR ESD?

A new era is opening again, as semiconductors transition from micro-electronics to nano-electronics with new questions, and uncertainties. As we enter the Nanoelectronic Era, the devices are getting smaller, but the interest in electrostatic discharge (ESD) phenomena is getting larger [1–5]. The semiconductor industry has invested forty years in the manufacturing of CMOS technology, but what about the future? One of the driving forces for this interest in ESD is the concern of the ability to manufacture these new nanostructures without destruction associated with static charge, and electrostatic discharge events. Coining a new field, "Nano ESD" is of both great global interest and concern as we enter the nanostructure world [1,6–9].

Will static charge impact the reliability, or the ability to manufacture nanostructures, or introduce them into the marketplace? Where are we going? Are there solutions to future ESD problems? Will ESD sensitivity manifest itself to be dependent on the type of structure and technology?

7.2 FACTORIES AND MANUFACTURING

In the future, factory materials and design may take a different course depending on the global expansion of factories and manufacturing facilities. The course that factories take will depend on its expectation of providing good ESD practices or factory costs. With globalization, the factory directions will be a balance between factory costs, the products that are produced and the value of ESD protection in the future.

ESD Basics: From Semiconductor Manufacturing to Product Use, First Edition. Steven H. Voldman.
© 2012 John Wiley & Sons, Ltd. Published 2012 by John Wiley & Sons, Ltd.

In manufacturing of nano-structures, the ESD sensitivity of the devices, components, and system products may play a key role in the direction of the factories, shipping and handling of future products. In the following sections, examples will be shown of different technologies, from photo-masks to nanowires. The chapter will first discuss examples of ESD failures where air gaps, or open surfaces are present in these devices, devices that are both either mechanically static, or mechanically dynamic structures. The focus of the chapter will include photo-masks, magnetic recording devices, and micro electromechanical machines (MEMs), but the subject is relevant to all electro-statically actuated and suspended structures (e.g., capacitors and inductors). The chapter will then discuss silicon devices, from bulk CMOS to silicon-on-insulator (SOI), and modern nanostructure FinFETs and nanowires.

7.3 PHOTO-MASKS AND RETICLES

Photo-masks are used in the formation of semiconductor devices. A photo-mask serves as the negative image of the desired structure to be created on the semiconductor wafer for printing purposes. A photo-sensitive material is formed on the wafer surface to produce the desired shape or structure. The photo-mask is used to expose the material, and leads to hardening of the photo-sensitive material. In photolithography, there is both positive tone and negative tone resist. Photo-masks are typically formed using a quartz substrate, with chrome shapes on the physical surface (Figure 7.1). The production of the photo-masks must be "defect free." If a defect occurs, this can lead to latent failure mechanisms to product failure.

Figure 7.1 Reticle

7.3.1 ESD Concerns in Photo-Masks

One of the key problems is the build up of electrostatic charge on the mask shapes. Between each chrome shape on the mask, a potential "spark gap" exists which can lead to electrical discharge when the electric potential exceeds the air breakdown. In the formation and handling of the photo-masks, charging issues can lead to damage of the chrome lines [10–19]. This occurs when a differential voltage is established across the photo-mask, both globally or locally. In the case of local damage, a differential voltage can be established between two adjacent chrome lines. The differential voltage can be dependent on the amount of charge build-up on each structure. If the physical regions are of different physical size, or shape, then the charge collection can be different, leading to a differential voltage. Electrical breakdown can occur as either a surface breakdown, or bulk air breakdown event.

Figure 7.2 shows an example of ESD discharge as a function of the spacing between two shapes on a photo-mask. In the figure, the failure damage indicates that the chrome material melted as a result of current flowing between the two physical shapes after electrical breakdown [4,6,10,19]. Figure 7.2a–e shows the electrical discharge as a function of the gap spacing (for the gap spacing of 1.5, 2, 2.5, 3 and 4 μm, respectively) [4]. At a gap spacing of 4 μm (Figure 7.2e, no damage is evident between the edge of the two chrome lines. At this gap spacing, there is evidence of onset of topography change on the upper corner of the line. In Figure 7.2d, at a gap spacing of 3 μm, there is little damage evident between the edge of the two chrome lines; in this case, there are changes in the topography of the end of the line

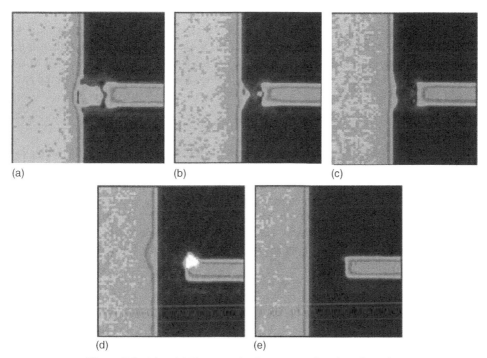

Figure 7.2   (a) to (e) Photo-masks damage as a function of spacing

and the vertical line. In Figure 7.2c, at a gap spacing of 2.5 μm, both corners and the end of the chrome line are actively involved in the discharge process. For smaller spacings, the damage level is evident in the gap between the vertical line and the end of the chrome line.

7.3.2 Avalanche Breakdown in Photo-Masks

Avalanche phenomenon is important to understand the breakdown process in air and other materials. Townsend, in 1915, noted that the breakdown occurs at a critical avalanche height,

$$H = e^{\alpha d} = \frac{1}{\gamma}$$

In this expression, the avalanche height H, is equal to the exponential of the product of the probability coefficient of ionization (number of ionizing impacts per electron and unit distance in the direction of the electric field) and electrode spacing. The avalanche height, H can also be expressed as the inverse of the probability coefficient of regeneration (number of new electrons released from the cathode per positive ion). The results of F. Paschen showed that avalanche breakdown process is a function of the product of the gas pressure and the distance between the electrodes. Paschen showed that

$$pd \approx \frac{d}{l}$$

where p is the pressure, d is the distance between the plates and l is the mean free path of the electrons. From the work of Paschen, a universal curve was established which followed the same characteristics independent of the gas in the gap. The Paschen curve is a plot of the logarithm of the breakdown voltage as a function of the logarithm of the product of the pressure and gap distance.

$$V_{BD} = f(pd)$$

At very low values of the p-d product, electrons must accelerate beyond the ionization limit to produce an avalanche process because the likelihood of impacts is too few. In this region, the breakdown voltage decreases with the increasing value of the pressure-gap product. This occurs until a minimum condition is reached. At very high values of the pressure-gap product, the number of inelastic collisions is higher and the breakdown voltage increases. This U-shaped dependence is characteristic of gas phenomenon. At high gas pressure, secondary processes, such as light emissions occur. In another form, in air,

$$p\lambda = 5 \times 10^{-3}$$

where p is in units of Torr, and λ is in centimeters. At a pressure of 1 atmosphere, the mean free path, λ, is 0.066 μm. The energy accumulated at one mean free path is 30 eV; this translates to the energy needed to initiate avalanche breakdown. Converting to the electric field, and voltage, this is equivalent to an electric field of E = 30 V/0.066 μm, or roughly

480 V/μm. Hence according to the Paschen theory, for breakdown in air (e.g., no surfaces) the breakdown voltage is less than 500 V for gap spacings less than 1 μm. Paschen showed that a breakdown relationship exists which is a function of the product of pressure and spacing. In our present day and future devices, the region of interest is in the U-shaped section of the Paschen curve where the breakdown voltage increases at smaller spacings. As we continue to make smaller line width for nanoelectronics, the spacing between lines is reduced, leading to electrostatic nanodischarges occurring between the mask shapes.

7.3.3 Electrical Model in Photo-Masks

The electrical response of these elements is also a function of the electrical parameters of the structure. An electrical model can be established to quantify the ESD event in photo-masks [10,19]. The event model can be depicted as a first and second chrome feature on a substrate. An air gap is between the two chrome features. The ground reference is the back of the reticle. The chrome features form a capacitor between the chrome shape and the back reference plane. The capacitance of the two chrome features can then be defined associated with the total area of the chrome feature, and the thickness of the quartz substrate wafer. The resistance of the chrome feature is a function of the feature sheet resistance, and the geometrical parameters. The inductance of the chrome line is associated with the inductance per unit length. The capacitor, C_1 and C_2, are the capacitors associated with the chrome feature and the substrate for the first and second chrome feature, respectively. The resistance and inductance are also the parameters for the two adjacent structures. In this representation, the capacitance and the arc resistance across the gap are not included in the model; this model assumption is valid for small gap regions, or when the gap capacitance and resistance is much smaller than the other capacitor and resistor terms. Note, in this representation, when the air gap conduction occurs, all the elements in the circuit are in a series configuration, forming an RLC response. J. Montoya, L. Levit, and A. Englisch showed that this representation could be fitted to the oscillation observed during electrostatic discharge [10,19]. The current in the discharge process can be represented as a decayed sinusoidal oscillation. The current is a function of the impedance. The impedance is the frequency times the sum of the two inductors (e.g., $Z = \omega\,(L_1 + L_2)$) [10,19].

$$I(t) = \frac{V}{\omega\{L_1 + L_2\}} e^{-\alpha t} \sin\{\omega t\}$$

The decay rate is associated with the equivalent R/L decay, where the equivalent resistance is the sum of the two resistors, and the equivalent inductance is the sum of the two inductors (e.g., $(R_1 + R_2)/(L_1 \mid L_2)$),

$$\alpha = \frac{1}{2}\left\{\frac{R_1 + R_2}{L_1 + L_2}\right\}$$

The frequency of oscillation can be obtained from the Kirchoff's voltage loop, with a characteristic oscillation [10,19],

$$\omega = \left\{ \alpha^2 + \left(\frac{1}{C_2} - \frac{1}{C_1} \right) \left(\frac{1}{L_1 + L_2} \right) \right\}$$

The response in the electrical discharge is associated with the sum of the two inductances, the sum of the two resistances, and the capacitance to the substrate ground plane. As a result, the RLC response is a function of the mask shapes (e.g., length and width), as well as the gap distance (Figure 7.3). The air gap distance also determines the voltage at which the break-down occurs.

7.3.4 Failure Defects in Photo-Masks

The discharge itself leads to melting of the chrome wires, which can lead to mask defects in the semiconductor chip. As the spacing decreases, these nanodefects are harder to observe in mask inspection. Figure 7.4 shows an atomic force microscope (AFM) image of chrome damage that can lead to latent defects. These nanodefects may prevent the ability to manu-facture structures without yield loss. Hence, it is of interest to address the manufacturing inspection process for future nano-structures.

Figure 7.5 shows the AFM image of the ESD defect between the two masks. As is evi-dent, the material from the mask exists between the two physical shapes.

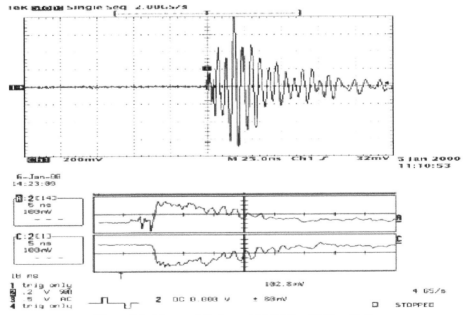

Figure 7.3 Measured discharge highlighting RLC oscillation

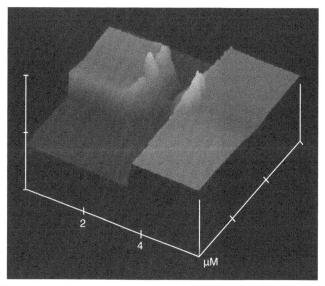

Figure 7.4 Atomic force microscope (AFM) image of latent defects in photomasks

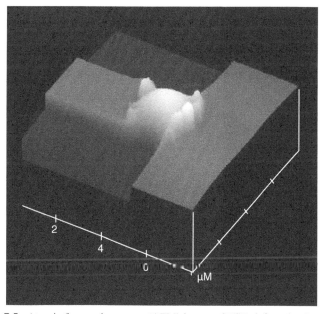

Figure 7.5 Atomic force microscope (AFM) image of ESD defects in photomasks

7.4 MAGNETIC RECORDING TECHNOLOGY

The magnetic recording industry uses a small thin film magneto-resistor (MR) to sense the magnetic field as the disk spins under the magneto-resistor mounted on a "magnetic head." To continue to scale down the size of disk drives, and to pack in as much areal density of information as possible, the MR industry continues to evolve the MR head device to smaller structures. The industry migrated from MR heads, to the giant magnetoresistors (GMR), and presently the tunneling magneto-resistor (TMR).

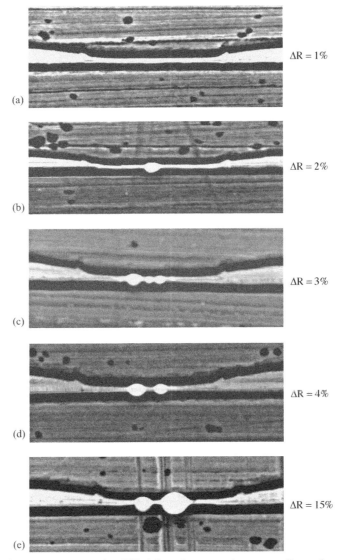

(a) $\Delta R = 1\%$

(b) $\Delta R = 2\%$

(c) $\Delta R = 3\%$

(d) $\Delta R = 4\%$

(e) $\Delta R = 15\%$

Figure 7.6 (a) to (e) MR stripe ESD failure damage and corresponding resistance shift

ESD and electromagnetic interference (EMI) has become a significant concern in the magnetic recording industry [4,20–28]. Today, the magnetic recording industry devices are the most ESD sensitive elements being manufactured; as a result, there is significant interest in how to manufacture a "Class 0" device. One of the primary reasons for low ESD robustness is there are no ESD devices or structures in hard disk drives to protect these elements.

In these devices, there is a significant number of electrical, magnetic, and aerodynamic failure mechanisms. There are also material changes of state, as well as electrical circuit states (e.g., initialization). Figure 7.6a–e shows an example of ESD failure damage in a magneto-resistor (MR) head structure as a function of resistance degradation shift [20]. As the defect changes in the MR stripe, the series resistance of the MR stripe changes. This impacts the magnetic characteristics of the MR film. The agglomeration of the MR device also changes the aerodynamic characteristics, as the MR stripe "flies" over the disk in the disk drive. Head-to-disk "crashes" can also occur, leading to disk drive failures.

Micro-breakdown can also occur between the magneto-resistor and adjacent shields, and substrates [20]. Along the surface, breakdown can occur, leading to damage of the MR stripe and the physical surface.

A second fundamental mechanism in MR heads is ESD failure between the MR stripe, and the adjacent magnetic shield structures. As in the photo-masks, electrostatic breakdown can occur across the surface between the two adjacent structures. In the MR head, the MR shields provide magnetic shielding from stray magnetic signal away from the MR stripe. As the signal levels decrease, the MR shields are placed closer to the MR stripe to enhance the MR stripe signal levels. Similar to the photo-mask, there is an air gap along the air-bearing surface (ABS), where the MR stripe and MR shields are separated at the air surface. Figure 7.7 shows an example of the ESD failure between the MR stripe, and the adjacent MR shields.

In this magnetic recording industry evolution, the size and film thickness are reduced to sense smaller signals; as a result, the human body model (HBM) ESD "robustness"

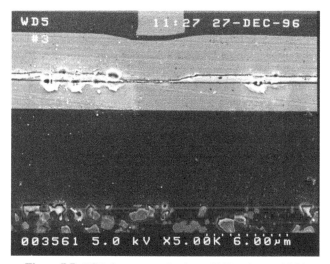

Figure 7.7 ESD damage mechanism of MR stripe to shield

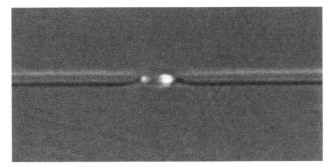

Figure 7.8 Tunneling magneto-resistor (TMR) ESD failure damage

decreased. In the 1993, the HBM ESD sensitivity of the MR head was 150 V HBM levels. With the introduction of the giant magneto-resistor (GMR) head, the HBM sensitivity level reduced to 35 V. This was followed by the introduction of the tunneling magneto-resistor (TMR) head, whose failure levels were less than 10 V HBM. How high can a human being be charged to? People have been measured at charged voltages as high as 35 kV. Semi-conductor components are designed typically to survive 2 kV (HBM) levels.

Today, the TMR is one of the most ESD sensitive nano-elements that is being manufactured [4,28]. The TMR device is teaching semiconductor manufacturing how to build, ship and assemble super-sensitive devices, leading the way for the nano-technology world that it will need in the future. Figure 7.8 shows the ESD damage of a TMR device [28].

7.5 MICRO-ELECTROMECHANICAL (MEM) DEVICES

Micro-electromechanical (MEM) structures are being developed for a large range of applications from basic electrical components, micro-engines, and micro-mirrors [4,7,29–38]. For motors, MEMs are being used as "energy scavengers" converting the energy of mechanical vibration to electrical energy. MEM electrical components include capacitors, inductors, and switches [4]. MEMs are also being used as arrays of micro-mirrors [35–38]. Today, there is a wide range of applications that are being explored, from electrical, biomedical, consumer, multi-media, hand-held electronics, energy to military needs. There are key issues common to all these MEMs:

- MEM structures contain air gaps.

- MEM structures contain air cavities or suspended structural elements.

- MEM structures have moveable elements.

- MEM structure mechanical motion is initiated by electrostatic fields (electro-statically actuated).

As a result, the failure mechanisms of MEM structures are a function of both electrical failure, and mechanical failure [4,7,29–38]. Mechanical failure can consist of creep, fatigue, wear, and "stiction." The failure of operation of a MEM structure can be associated with structural damage impeding the motion of the move-able elements. Electrical failure can be associated with surface or air gap breakdown. The electrical breakdown can lead to structural melting, electrostatic deflection, or damage to the structure; this may manifest itself as associated with resistance, capacitance or inductive change in the operation of a MEM device. Electrical failure can be related to changes in the device dc or ac parameters or leakage current as well. Today, not all ESD failure mechanisms have been demonstrated in all MEM application spaces. Much of the observations and knowledge of these MEM ESD failures can be transferred to other MEM applications, such as MEM inductors, and MEM capacitors [4].

MEMs pose a new challenge due to the fact that many of these nano-elements are electro-statically actuated elements. How does one avoid electrostatic discharge problems in an electro-statically actuated element? How does one avoid electrostatic discharge in the air gaps, and suspended structures? As in the photo-mask, electrical charging can occur between two closely spaced elements. In MEM structures, segments of the elements are closely spaced, with a gap between to provide the electrostatic actuation. As a result, an electrical spark can occur in the gap leading to melting of the component, and at times causing them to "stick" together. In these MEM structures, both functional impacts, and "stiction" can occur.

7.5.1 ESD Concerns in Micro-electromechanical (MEM) Devices

In electro-statically actuated devices, ESD failure is a function of a number of issues. A first failure mechanism is electrical breakdown. Electrical breakdown can occur across the air gap between the actuation source and the physical structure or membrane in the device. As discussed previously, the Paschen curve discusses the relationship between the breakdown voltage and the gap size. An interesting feature of the electrostatically actuated devices is that as the mechanical motion occurs, the gap size varies during operation. A unique difference between field emission devices (FED), spark gaps, magnetic recording heads (e.g., AMR, GMR, TMR, and colossus magneto-resistor (CMR), and photo-mask structures is that in the MEM structure, the gap is variable. This gap variation is dependent on the following:

- Electrical state (e.g., actuated, non-actuated or in a switch transition state).

- Residual charge (pre- and post-ESD events).

- Mechanical deformation (e.g., pre-strained condition).

- Electric field (pre- and post-ESD events).

For the electrical state, the structure can be "open", "closed" or in a switch transition state. In this case, the gap dimension will be different. In the case that the gap space is small, material displacement can lead to "stiction." Melting between the two surfaces can create a merging of the two structures preventing operation, also known as "stiction." In photo-masks, it serves as a defect; in MEM structures, it prevents operational function of the MEM device. The mechanical "spring constant" of the structure also influences whether or not the structure restores itself after discharge (e.g., the stiffness of the structure influences the return response).

Residual charge can lead to changes in the "initial state" of the structure, influencing the gap spacing prior to an ESD event. This "pre-charge" produces an electric field, inducing displacement or deflection of the structure. Hence pre-conditioned states (both electrical and mechanical) lead to smaller gap spacings as well as an initial electric field, which will influence the HBM, machine model (MM) and transmission line pulse (TLP) results.

Mechanical deformation, due to pre-strain conditions or non-elastic deformation, can also lead to change in the position of the actuated device. Hence the mechanical initial state leads to smaller gap spacings as well as an initial electric field, which will influence the HBM, MM and TLP results.

7.6 MICRO-MOTORS

Micro-mechanical engines are valuable in the future in a wide range of applications [29,30]. An example of a micro-engine contains gears and linear actuators. The micro-engine has a rotating gear. The gear rotates around a hub assembly which is attached to the substrate. The rotating gear is mechanically connected to linear actuators. The micro-machine has an orthogonal comb drive; these "comb drives" contain a first set of stationary comb fingers, and a second set of grounded comb fingers. The first set is a stationary set of comb fingers which is electrically insulated from electrical ground. The second comb fingers is a grounded comb mechanically attached to a moveable shuttle structure. Springs suspend the shuttle above the ground plane and serve as an electrical ground potential.

7.6.1 ESD Concerns in Micro-Motors

J. Walraven of Sandia laboratories first addressed ESD damage concerns in a micro-motor [29,30]. Figure 7.9 shows an example of a torsional ratcheted actuator (TRA) (developed by Sandia National Laboratories). Sandia National Laboratories have been exploring the reliability of these motors from an electrostatic perspective. Electrostatic discharge testing was completed to evaluate the failure mechanisms. J. Walraven noted the existence of damage to the gear rotation and "stiction" of the motor elements. "Nano-welding" was observed from the current of the ESD event. Another form of reliability is the residual particles that are left after a discharge process can interfere with mechanical rotation of elements.

J. Walraven *et al.* showed that these actuators had an HBM ESD failure level of 100 to 130 V (HBM) [29,30]. Additionally, the MM ESD failure levels were 95 to 120 V (MM) [29,30] (Figure 7.10). In semiconductor devices, typically the ratio of HBM-to-MM ESD results is between 5:1 to 20:1. In semiconductor devices, this failure ratio is very dependent on current density, and temperature. But in these micro-engines, the failures are associated with the breakdown voltage across air gaps. Hence, the failures are more similar to spark gaps, photo-masks, and magnetic recording devices than ESD failures observed in semiconductor components.

Failure analysis showed that the ESD failure mechanism is associated with a second layer polysilicon comb fingers adhering to a first layer polysilicon ground plane. An electrical

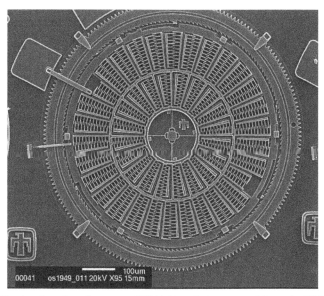

Figure 7.9   Torsional ratcheted actuator (TRA). Permission granted from the ESD Association

discharge occurred between these two polysilicon layers, leading to molten polysilicon at the region of failure. The electrical failure signature was determined by TIVA analysis. TIVA analysis was verified with SEM analysis; ESD failure occurred at the polysilicon comb finger. Charged device model (CDM) testing of these structures demonstrated that the failure levels exceeded 1000 V (CDM); some actuators exceeded levels over 2000 V (CDM) with no evidence of failure [29]. In these micro-engines, the operational structures are not contained

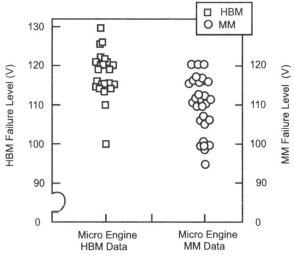

Figure 7.10   Torsional ratcheted actuator (TRA) HBM and MM ESD results

within the silicon substrate; these structures are electrically isolated from the substrate wafers, and are less vulnerable to bulk-silicon semiconductor devices. Failure analysis verified that no damage was observed between the fixed or moveable comb fingers to the substrate. In some cases, the polysilicon damage was observed on the polysilicon comb structures; in cases where there was no adherence, the micro-engine was still functional. Hence, evidence shows that in some structures, although there is silicon or polysilicon damage, if the damage does not lead to physical adherence, the micro-engine can still maintain functional operation. But, if the discharge process leads to adhesion or "welding" of the two physical structures, the micro-engine cannot operate [29,30].

7.7 MICRO-ELECTROMECHANICAL (MEM) RF SWITCHES

For RF applications, RF MEM switches have advantages over conventional switches. For the electrical state, the structure can be "open", "closed" or in a switch transition state. In these cases, the gap dimension will be different. ESD can be a concern in these electrostatically actuated switches [31–34]. In cases where the gap space is small, material displacement can lead to "stiction." Melting between the two surfaces can create a merging of the two structures preventing operation, also known as "stiction." In photo-masks, it serves as a defect; in MEM structures, it prevents operational function of the MEM device.).

7.7.1 ESD Concerns in Micro-electromechanical (MEM) RF Switches

A. Tazzoli studied the impact of ESD on RF MEM switch devices [31,34]. As will be noted by Tazzoli [31,34], the "spring constant" of the structure also influences whether or not the structure restores itself after discharge (e.g., the stiffness of the structure influences the return response). Mechanical deformation, due to pre-strain conditions or non-elastic deformation, can also lead to change in the position of the actuated device. Hence the mechanical initial state leads to smaller gap spacings as well as an initial electric field, which will influence the ESD results.

A second issue is the mechanical failure of structural elements (e.g., broken structure). In the case of structural elements, mechanical deformation can be evaluated using solid mechanics "beam theory." A structural element can be regarded as a beam structure with a given cross sectional area. An example of understanding the beam deflection is a function of the support. For the example of a cantilever structure, it can be assumed that a force exists that is distributed across the beam length, where the force pulls the beam toward a planar surface. An electric field can be formed between the beam structure and the surface, leading to the applied force on the beam structure. The beam displacement is associated with the magnitude of the force, the beam length, the beam thickness, and the Young's modulus of the beam material. The magnitude of the force is a function of the gap size between the beam and the surface (e.g., defining the electric field established).

A model used for MEM structures to evaluate beam displacement is the Osterberg model. The model evaluates the voltage needed to have the end of the cantilever element to touch the surface. This voltage is known as the collapse voltage. The collapse voltage where the beam deflects to the surface is

$$V_c = \left\{ \frac{16Et^3 g_O^3}{81\varepsilon l^4} \right\}^{\frac{1}{2}}$$

where E is the Young's modulus (in units of mega-Pascals), t is the beam thickness, g is the gap between the cantilever beam and the surface, ε is the permittivity of air, and l is the length of the beam. Broken physical elements can lead to residual materials within the air gap, influencing functional operation or electrical shorting.

Figure 7.11 shows an example of an RF switch, with an RF "input" (RF(IN)), an RF "output" (RF(OUT)), and an actuating bridge structure. In these RF MEM switches, ESD events influence the mechanical motion as well as influencing the RF S-parameters [31,34]. Figure 7.12 shows the damage of the RF MEM switch after ESD stress. The ESD damage can manifest itself between the RF (IN) and RF (OUT), as well as being an actuator to RF input or output signals. Additionally, the ESD current can also lead to "nano-welding" of the elements in the switch element, impacting its functionality, as was observed in micro-motors.

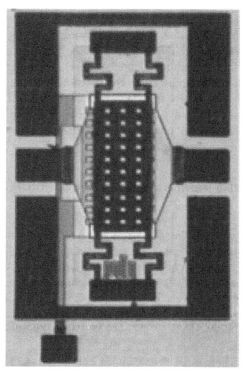

Figure 7.11 Radio frequency (RF) switch

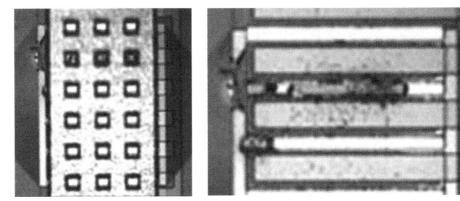

Figure 7.12 Radio frequency (RF) MEM damage after an ESD event. Permission granted from the ESD Association

7.8 MICRO-ELECTROMECHANICAL (MEM) MIRRORS

Micro-mirrors have present day and future applications in systems. Optical MEM systems contain electrostatically-actuated micro-mirror structures [35–38]. These micro-mirror elements can be used in a wide range of applications that utilize coherent or incoherent light. For coherent light applications that utilize lasers, the MEM structure can be used for read heads in disk drives, laser printers, bar code readers, and scanning machines. As a result, for hand-held or non-portable applications, these micro-mirror systems have significant value. Using an electrostatic actuator, the micro-mirror can be tilted by the capacitive coupling between the mirror and the pad on the silicon substrate. In these structures, the actuation voltage influences the tilt angle; the tilt angle is a function of the actuation voltage. As was shown in the other MEM applications, ESD failures can occur. In micro-mirrors, the uniqueness is that the deflection and bending of the element is critical to the application. In the MEM switch, it is influenced by the "open" or "closed" state of the switch; hence is digitized whereas the MEM micro-mirror is "analog" in nature.

7.8.1 ESD Concerns in Micro-electromechanical (MEM) Mirrors

Figure 7.13 is an SEM of the micro-mirror array after ESD testing. A mirror can be one of the plates of a capacitor element. Micro-mirrors can have a supporting beam in the center and a "left and right" capacitor plate on each side. By applying an electric field, the electrostatic attraction can lead to a tilting of a mirror element, like a "see-saw." The tilt angle is a function of the applied electric field. Micro-mirrors also exhibited ESD concerns. ESD events can lead to damage between the mirror and the actuator; this can lead to distortion of the tilt angle and rotation of the micro-mirror structures. In an array of micro-mirrors, mirrors are separated by small gaps. ESD events can occur between the mirrors due to discharge events, leading to damage of the mirror surface [38].

As in the RF MEM switches, the ESD damage can impact the mechanical motion, and system response. In the micro-mirror array, ESD damage leads to reduction of the tilt angle

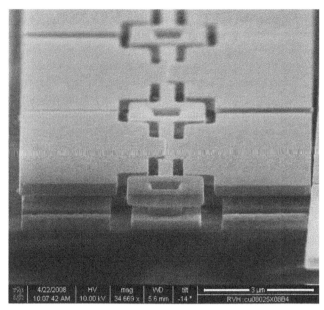

Figure 7.13 Micro-mirror structure

of the structure. As the actuation voltage was increased, tilting of the micro-mirror still occurred, but the range of tilt angle was significantly impacted by the magnitude of the ESD event (e.g., impacting its functionality). HBM ESD testing was performed for different size micro-mirrors, in both single and array configuration. It was noted by S. Sangameswaran *et al.*, that one of the failure mechanisms in a micro-array is associated with failures between the adjacent micro-mirrors [38]. Between adjacent mirrors, an air gap exists, forming a "spark gap," which can lead to mirror-to-mirror ESD failures.

7.9 TRANSISTORS

In silicon technology, there have been both evolutionary and revolutionary transitions in the semiconductor transistor as it has been scaled to smaller dimensions. The semiconductor dimensional scaling has gone from the micro-meters to nano-meters over the last thirty years; where MOSFET constant electric field scaling has influenced the ESD robustness of the transistor [1–5]. In the semiconductor device, there has been transition in the semiconductor wafer from low-doping concentration, to high doping, and once again to low doped wafer [39,40]. In well design, technology evolved from single tub, to dual-well, to triple-well; this evolution was influenced by semiconductor process equipment (e.g., MeV implantation), to semiconductor device requirements, isolation (e.g., latchup requirements), and device offerings (e.g., low voltage and high voltage devices). In isolation technology, it has evolved from recessed oxide (ROX), to local oxidation (LOCOS), and presently shallow trench isolation (STI) to provide improved topography and dimensional scaling [39–42].

In the transistor junctions, technology evolved from single implants, to L_{DD}, to extension implants. Today, using silicon germanium (SiGe), strain engineering can be applied to change the mobility by introducing regions of tensile, or compressive states. With mechanical strain, the spacing of the lattice atoms can be modified to vary the collision frequency of free carriers with the lattice atoms. Mechanical strain can also be introduced using films over the gate structure that introduce compression and tension along the silicon surface.

The interconnect system must also scale with the silicon transistor due to integrate the interconnects to the MOSFET device; to achieve this, innovation in the planarity, and metallurgy was introduced. Planarity was achieved using chemical mechanical polishing (CMP). In the metallurgy, revolutionary changes occurred in the metal films and via contact structures, by transitioning from aluminum to copper-based interconnects, where this transition changed the ESD robustness of semiconductors [43,44]. For semiconductor chip performance, low-k dielectrics were introduced to lower the capacitance and line-to-line coupling of the interconnects; this also influenced the ESD robustness of the interconnects [45]. Today, through silicon vias (TSV) have been introduced for multi-chip modules, stacked chips, and increased functional bandwidth.

7.9.1 Transistors – Bulk vs. SOI Technology

To achieve continued performance objectives, and remain on the Moore's Law curve, it was believed that MOSFET junction capacitance was a performance impediment. Partially depleted silicon on insulator (PD-SOI) was introduced to serve as a natural evolutionary change from the "bulk CMOS" transistor. A concern with SOI technology was the ability to achieve ESD protection and migrate bulk CMOS applications into SOI applications, without degradation of the ESD robustness of products [1–5]. ESD development began in 1991 to demonstrate the ability to offer SOI as a mainstream replacement for bulk CMOS technology. Figure 7.14 shows an example of the bulk CMOS to SOI remap process. After years of

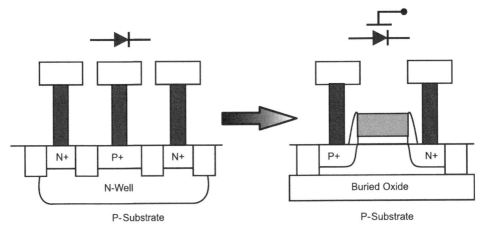

Figure 7.14 Bulk to SOI Mapping

work, it was found that it is possible to provide ESD robust products in SOI [46,47]. This work opened the door for semiconductor development, and provided a dilemma as well. The question for technology was whether they choose the path of bulk CMOS or SOI technology for high performance, advanced technology. As we migrate to 22 nm and sub-22 nm technology, both are being explored as the candidates for advanced CMOS.

7.9.2 Transistors and FinFETs

As the silicon transistor scales from a micron technology world, to the nanodevice world, it is believed that semiconductor devices must achieve two paradigm shifts: first, introduce fully-depleted transistors, and second, extend into the third dimension.

In semiconductors, in order to pack more transistors into more space, and to provide higher performance, the transistor is leaving the two dimensional (2-D) wafer and becoming more three dimensional (3-D). New 3-D devices known as "FinFETs" (also referred to as multiple gate field effect transistors (MUGFETs)) are being proposed for future CMOS technologies [48–50]. The name "FinFET" comes from the 3-D vertical elongated "fin" structure of silicon from the wafer surface. Multiple gate and wrap-around gate structures are of interest for advanced development of the desire to increase the current conduction in a MOSFET transistor per unit area. For density improvements, exploratory work in the 45 nm and 32 nm technology generations has been initiated. In the 22 nm or a future SOI technology, the possibility of using the fully-depleted SOI transistors and SOI FinFET transistor is very likely.

Today, these transistors are being demonstrated in 45 nm technology with potential usage in the future. In these structures, the MOSFET gate "wraps around" all sides of the "fin." Each "fin" is physically isolated from the adjacent fins with a common gate extending over all fins perpendicular to the flow of the current.

Figure 7.15 shows an example of a FinFET structure. Instead of a planar device surface, the FinFET is segmented into vertical pillars of silicon. The MOSFET gate wraps around the "fin" structure on three sides, meandering up and down over the parallel fin structures. The FinFET consists of parallel nanochannels to conduct the MOSFET current.

7.9.3 ESD in FinFETs

ESD measurements have been reported in these FinFET transistors, providing new challenges on how to design and optimize the new transistors [39,40]. From a nano-structure perspective, the nano-channels influence the "electro-striction" which occurs in a planar device, quantizing the current constriction into a number of parallel channels.

ESD failure of FinFET structures is of interest due to the inter-relationship of the FIN width, and the nature of current electro-constriction [1]. In planar MOSFETs, the lateral current constriction across the MOSFET width has prevented the ability to predict the width scaling of the MOSFET structure during a high current state. In the FinFET structure, the FinFET is significantly smaller than the current constriction, leading to the current being distributed across multiple fins.

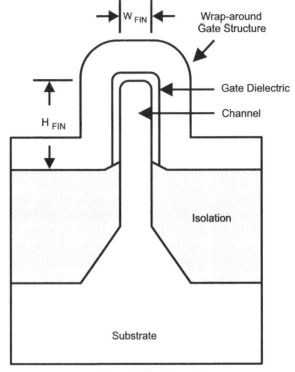

Figure 7.15 FinFETS

C. Russ *et al.* demonstrated the first experimental work of ESD failure of a SOI FinFET structure. TLP I-V characteristics of 320 parallel fins for 150 and 90 nm length FIN structures that have a fin width of 50 nm were evaluated [48]. A second study shows a second TLP I-V characteristic of 500 parallel fins for 250 and 120 nm length. ESD failure occurs at the FinFET TLP I-V I_{t2} current level. In the FinFET structure, at the ESD failure level, many of the FinFET channels are displaced as a result of the ESD event. Note that the ESD failure occurs in adjacent FinFET channels, and are not randomly distributed through the structure (Figure 7.16).

These FinFET structures can be used in a diode configuration as well as a MOSFET configuration [48]. TLP I-V characteristics were compared for a planar SOI lateral diode versus a SOI FinFET structure. The response of the TLP FinFET diode structure was similar to the planar diode structure. In this investigation, the planar SOI lateral diode had a lower resistance compared to the SOI FinFET diode structure [48]. In the diode configured FinFET structure, at the ESD failure level, many of the diode SOI FinFET channels are displaced as a result of the ESD event. Note that, as was shown in the prior MOSFET structure, ESD failure occurs in adjacent FinFET channels, and is not randomly distributed through the structure.

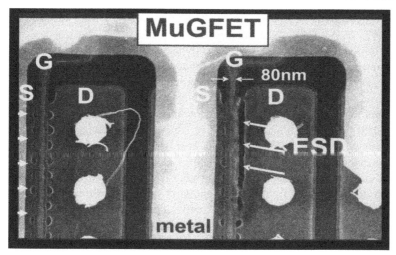

Figure 7.16 ESD damage in FinFET device. Permission granted from the ESD Association

7.10 SILICON NANOWIRES

As semiconductor devices continue to scale, there is interest in nanowires for both current conduction, and switches. Nanowire (NW) structures may be used in future applications in the sub-20 nm technology generation. Silicon nano-wires have been demonstrated with 15 nm gate length, and 4 nm radius [51]. The silicon NW can serve as a MOSFET structure that is fully depleted due to the MOSFET gate structure surrounding the nanowire on all sides [51].

A roadblock to practical implementations of these structures is the ability to survive ESD events. Experimental work has already begun to understand the ESD robustness of the silicon nanowire, with the first reported ESD results being published in September 2009, by W. Liu, J.J. Liou, A. Chung, Y.H. Jeong, W.C. Chen, and H.C. Lin [52]. TLP testing was completed on NW structures that were varied in the number of parallel wires, and the nanowire channel length. First, ESD results increased with the number of parallel nanowires. Second, as the nanowire channel length decreased, the TLP current increased. Using 100 parallel nanowires, the critical current-to-failure of the structure occurred at 20 mA. For 70 parallel channels, as the channel length decreased from 2 to 0.4 μm, the critical current-to-failure increased from 7 to 18 mA [52].

7.11 CARBON NANOTUBES

Carbon nanotubes (CNT) are of significant interest to the nanostructure world [1,56,57]. CNTs are being proposed for future electronic applications, with a significant amount of research and development. CNTs may serve as circuitry on both the periphery and interior of circuit applications. Some of the questions to be asked are how do we protect CNT circuitry? How do we build ESD protection circuits out of CNTs? Can we conduct the current using

Carbon nanowires ? Will they have significant value to construct ESD networks? Is there a concern today?

A reason for the ESD concern is to make sure that electrostatics do not interfere with the mainstream introduction of nanotubes into the manufacturing and the marketplace. If nanotubes and nanowires are to be a reality in future products, we must be able to establish practical techniques for manufacture processing, shipping and handling of these components.

The reason for the interest is also the great potential of nanowires in their current-carrying capabilities. If we can learn how to harness the nanowires' advantages, tremendous advantages may be only ahead of us. The search for ever faster, lower power consuming and thus smaller electronic devices is leading to structures having geometries of width and length smaller than current semiconductor image forming technology can generate. Of particular interest are structures formed from carbon-nanotubes such as quantum dots, CNT wires and CNT field effect transistors (CNTFETs). Circuits constructed using CNTFETs and CNT wires will require ESD protection and structures and methods for providing such protection are virtually unknown. Accordingly, there exists a need for compatible CNT diodes and ESD circuits for protecting CNT based electronic devices.

Carbon nanotubes are closed-cage molecules composed of sp^2-hybridized carbon atoms arranged in hexagons and pentagons. Carbon nanotubes come in two types, single-wall nanotubes, which are hollow tube like structures or and multi-wall nanotubes. Multi-wall carbon nanotubes resemble sets of concentric cylinders. Single-wall and multi-wall domain carbon nanotubes produce quantized conduction characteristics which are significantly different from standard electronics. The breakdown characteristics provide unique behavior which will be of interest from functionality, reliability and ESD perspectives. Using two doped CNTs, a p-n diode can be formed to serve as a protection device. As noted by Jia Chen of IBM T.J. Watson Research, orientation and physical area of the CNTs will influence the capability of the cross-point CNT [54]. For example, a CNT diode includes a first CNT extending in a first lengthwise direction from a catalytic nano-particle and electrically contacted by a metal contact at an end opposite the end in contact with the catalytic nano-particle. CNT diode also includes a second CNT extending in a second lengthwise direction from a catalytic nano-particle and electrically contacted by a metal contact at an end opposite the end in contact with the catalytic nano-particle. CNTs can lie in two different but parallel planes that are both parallel to a plane defined by a top surface of an insulating layer. Catalytic nano-particles are, in one example, iron/cobalt (Fe/Co) or iron/cobalt/nickel (Fe/Co/Ni) when CNTs are formed by chemical vapor deposition (CVD) from carbon containing gases, e.g., a carbon monoxide (CO) and hydrogen mixture.

Today, there still remains an open question of the future of CNT ESD robustness, and whether ESD will be an impediment to CNT integration into modern day electronic systems. Hence, it is not clear at this time whether ESD robustness of the CNT is a future advantage or disadvantage. There is significantly more work ahead to answer this question.

7.12 FUTURE SYSTEMS AND SYSTEM DESIGNS

In the last few years, there has been rapid change in the electronic industry. Large servers have been migrated to small servers. Storage of information has been migrating from the workplace to "clouds" and cloud centers. Desktop computers are being replaced with laptops.

Laptops are eliminating disk drives and CD/DVD ports. Laptops are being reduced in size from 17 inch screens to 11 inch screens. Laptops are becoming thinner; from thick to ultra-thin designs. Laptop encasements are being reduced to single machined units, with only a minimal of exposed ports. And, laptops are being replaced with tablets.

Tablets are being integrated with clouds, computers, and smart phones. Tablets have "touch screens," eliminating entry points for EMI and EMC issues. This year, tablets are starting to be reduced in size from the laptop scale, to smaller, lighter tablets; almost approaching the size of large cell phones!

Telephone "home lines" are being eliminated and replaced with wireless cell phone lines. Cell phones are migrating from keyboard cell phones, to "touch screen" cell phones with a minimal number of exposed ports, with the same trend as the tablets.

In electronic trade shows, "touch less" systems are being introduced, where the user does not even have to touch the touch screen but "sweep" near the surface to advance the screens. With the introduction of touch-less systems, there will be a shift from ESD and EOS issues to more focus on EMI and EMC issues.

Today, CMOS technology is migrating from planar transistors, to "Fin" transistors, and silicon on insulator (SOI) technology, eliminating parasitic devices and latchup concerns.

With the rapid changes in systems, and with evolutionary and revolutionary changes in systems, components and devices, the importance of ESD, EOS, latchup, EMI and EMC will be influenced. Some of these issues may be less important, and other phenomena may increase in importance. This trend is already happening today andwith the changes made today, in the future, this will be also true.

7.13 SUMMARY AND CLOSING COMMENTS

As the dimensional scaling of devices approaches the nano-meter dimensions, all devices will have to address the implications of static charge, electrostatic discharge (ESD), electro-magnetic interference (EMI), and electrical overstress (EOS). This will be true in photo-masks, magnetic recording devices, semiconductor devices, nanowires and nanotubes.

In photo-masks, magnetic recording devices, and MEM structures, there are unique failures which occur in these structures due to the presence of gaps, cavities, and floating structures. These structures in many cases are mechanically static, such as photo-masks, where others are electro-statically actuated. It was shown that the ESD failure is associated with the voltage of the structure exceeding the functional electro-static actuation voltage; this can lead to melting, or fusing of moveable elements, mechanical failure, and damage to the device. In the case of mechanically static devices, such as photo-masks, the residual material impacts the imaging process. In the case of mechanically dynamic devices, the residuals can lead to failure of mechanical operation. Given the yield stress of the moving structure is exceeded, mechanical damage occurs and particle elements and broken elements ensue. A key discovery in these structures is that there is little difference between the HBM and MM results; in air gap based failure mechanisms, the failure is associated with the voltage break-down of the air gap; this is very different from silicon-based diode and MOSFET structures.

Additionally, in many of the mechanically actuated devices, they are electrically isolated from the supporting substrate, leading to high CDM voltage levels. Again, this is very distinct from the silicon-based device elements, and is more akin to ESD failures observed in MR heads and photo-masks. Additionally, for very short pulses, the ESD pulse width is significantly smaller than the spring constant response time of the mechanical structures. It is also clear that the failures are associated with voltage levels that exceed the functional operational voltage; MEM structures are vulnerable to voltages outside of the functional regime which can lead to air breakdown, or mechanical deflections outside of the allowed range of both electromechanical operations.

The challenge to silicon devices is how to maintain robustness of semiconductor components as performance objectives continue to increase. As can be seen, the migration to nano-structures may not be as difficult as anticipated based on the early experimental results on silicon nanowires.

In summary, if the nano-structure is small, and exposed to external sources, then it is a candidate for a concern associated with static charge affects. In the future, there will be a need for research and development of electrostatic breakdown for present day and future nano-structures. In addition to the ESD issue, there are electromagnetic interference (EMI) concerns on both a component or system level. Electrical and magnetic fields are generated by the discharge process, which can also be a concern for future systems that contain nano-structures. Hence, there is a need to evaluate not only the nano-structure semiconductor devices and components, but "Nano-ESD" must be addressed at the system level (i.e., "System level Nano-ESD"). Future systems that contain these nano-structures, must also be evaluated to quantify the system level ESD concerns.

Will there be ESD solutions in the future for these nano-structures? In the past, where ESD appeared to be a roadblock, somehow, there was a solution. It will be of interest to see how the engineering world will address Nano-ESD. Today, the question remains open and unanswered [55].

REFERENCES

1. Voldman, S. (2004) *ESD: Physics and Devices*, John Wiley and Sons, Ltd., Chichester, England.
2. Voldman, S. (2005) *ESD: Circuits and Devices*, John Wiley and Sons, Ltd., Chichester, England.
3. Voldman, S. (2006) *ESD: RF Technology and Circuits*, John Wiley and Sons, Ltd., Chichester, England.
4. Voldman, S. (2009) *ESD: Failure Mechanisms and Models*, John Wiley and Sons, Ltd., Chichester, England.
5. Voldman, S. (1998) The impact of MOSFET technology evolution and scaling on electrostatic discharge protection. *Microelectronics Reliability*, **38**, 1649–1668.
6. Voldman, S. (2002) Lightning rods for nanoelectronics. *Scientific American*, **287** (4), 90–97.
7. Voldman, S. (2009) Nano Electrostatic Discharge (ESD). *IEEE Nano Technology Magazine*, **3** (3), 12–15.
8. Voldman, S. (2006) Electrostatic discharge protection in the nano-technology era – Will we be able to provide ESD protection in the future? Proceedings of the International Conference on Semiconductors and Integrated Circuit Technology (ICSICT), Shanghai, China, October 2006.

9. Voldman, S. (October 12–15 2007) Electrostatic discharge in the nano-technology era, Keynote Talk. Application Specific Circuits and Networks (ASICON) 2007, Guilin, China.

10. Montoya, J., Levit, L., and Englisch, A. (2000) A study of the mechanisms for ESD damage in reticles. Proceedings of the Electrical Overstress/Electrostatic Discharge (EOS/ESD) Symposium, pp. 394–405.

11. Greig, E. (1995) Controlling static charge in photolithography areas. *Micro Magazine*, **13** (5), 33–38.

12. Steinman, A. and Montoya, J.A. (1997) Developing an exit charge specification for semiconductor production equipment. *Micro Magazine*, **15** (4), 32–39.

13. Levit, LB, and Menear, J. (1998) Measuring and quantifying static charge in cleanrooms and process tools. *Solid State Technology*, **41** (2), 85–92.

14. Wiley, J. and Steinman, A. (April 1999) Ultrapure materials-reticles: Investigating a new generation of ESD-induced reticle defect, *Micro Magazine*, www.micromagazine.com.

15. Wang, K.C. (2000) ESD and reticle damage. International SEMATECH Electronic Discharge Impact and Control Workshop, October 9th, 2000, SEMATECH, Austin, Texas.

16. Steinman, A. (2000) SEMI E78-0998 Electrostatic Compatibility. International SEMATECH Electronic Discharge Impact and Control Workshop, October 9th, 2000, SEMATECH, Austin, Texas.

17. Armentrout, L. (2000) Analysis of ESD/reticle SMIF pods. International SEMATECH Electronic Discharge Impact and Control Workshop, October 9th, 2000, SEMATECH, Austin, Texas.

18. Pendley, M. (2000) ESD induced EMI Detection Techniques. International SEMATECH Electronic Discharge Impact and Control Workshop, October 9th, 2000, SEMATECH, Austin, Texas.

19. Montoya, J., Levit, L., and Englisch, A. (2001) A study of the mechanisms for ESD Damage in reticles. *IEEE Transactions on Electronic Packaging and Manufacturing*, **24** (2), 78–85.

20. Wallash, A.J., Hughbanks, T., and Voldman, S. (1995) ESD failure mechanisms of inductive and magnetoresistive recording heads. Proceedings of the Electrical Overstress/Electrostatic Discharge (EOS/ESD) Symposium, pp. 322–330.

21. Tian, H. and Lee, J.K. (1995) Electrostatic discharge damage of MR heads. Proceedings of INTERMAG 1995.

22. Tian, H. and Lee, J.K. (1995) Electrostatic discharge damage of MR Heads. *IEEE Transactions of Magnetics*, **31** (6), 2624–2626.

23. Cheung, T. and Rice, A. (1996) An Investigation of ESD protection for magnetoresistive heads. Proceedings of the Electrical Overstress/Electrostatic Discharge (EOS/ESD) Symposium, pp. 1–7.

24. Lam, C., Chang, C., and Karimi, R. (1998) A study of ESD sensitivity of AMR and GMR recording heads. Proceedings of the Electrical Overstress/Electrostatic Discharge (EOS/ESD) Symposium, pp. 360–367.

25. Wallash, A. and Smith, D. (1998) Electromagnetic interference (EMI) damage to giant magnetoresistive (GMR) recording heads. Proceedings of the Electrical Overstress/Electrostatic Discharge (EOS/ESD) Symposium, pp. 332–340.

26. Wallash, A. and Kim, Y.K. (1998) Magnetic changes in GMR heads caused by electrostatic discharge. *IEEE Transactions of Magnetics*, **34** (4), 1519–1521.

27. Chen, T.W., Wallash, A.J., and Dutton, R. (2008) Ultra-fast transmission line pulse testing of tunneling and giant magnetoresistor heads. Proceedings of the Electrical Overstress/Electrostatic Discharge (EOS/ESD) Symposium, pp. 258–261.

28. Wallash, A. (2003) ESD challenges in magnetic recording: Past, present and future. Proceedings of the International Reliability Physics Symposium (IRPS), pp. 222–228.

29. Walraven, J.A., Soden, J.M., Tanner, D.M. *et al.* (2000) Electrostatic discharge/electrical overstress susceptibility in MEMs: A new failure mode. Proceedings of the Society of Photo-Optical Instrumentation Engineers (SPIE), vol. 4180 pp. 30–39.

30. Walraven, J.A., Soden, J.M., Cole, E.I. *et al.* (2001) Human body model, machine model, and charged device model ESD testing of surface micromachined microelectromechanical Systems (MEMS). Proceedings of the Electrical Overstress/Electrostatic Discharge (EOS/ESD) Symposium, pp. 238–247.

31. Tazzoli, A., Peretti, V., Zanoni, E., and Meneghesso, G. (2006) Transmission line pulse (TLP) testing of radio frequency (RF) micro-machined micro-electromechanical systems (MEMS). Proceedings of the Electrical Overstress/Electrostatic Discharge (EOS/ESD) Symposium, pp. 295–303.

32. Ruan, J., Nolhier, N., Bafluer, M. *et al.* (2007) Electrostatic discharge failure analysis of capacitive RF MEM switches. *Microelectronic Reliability Journal*, vol. **47**, 1818–1822.

33. Tazzoli, A., Peretti, V., and Meneghesso, G. (2007) Electrostatic discharge and cycling effects on ohmic and capacitive RF-MEMS switches. *IEEE Transactions on Device and Material Reliability (TDMR)*, **7** (3), 429–437.

34. Tazzoli, A., Peretti, V., Autuzi, E., and Meneghesso, G. (2008) EOS/ESD sensitivity of functional RF MEMS switches. Proceedings of the Electrical Overstress/Electrostatic Discharge (EOS/ESD) Symposium, pp. 272–280.

35. Zhang, X.M., Chau, F.S., Quan, C., and Liu, A.Q. (1999) Modeling of optical torsion micro-mirror. Proceedings of the Society of Photo-Optical Instrumentation Engineers (SPIE), Vol. 3899, pp. 109–116.

36. Douglass, M.R. (1998) Lifetime estimates and unique failure mechanisms of the digital micro-mirror device (DMD). Proceedings of the International Reliability Physics Symposium (IRPS), pp. 9–16.

37. Gromova, M., Haspeslagh, L., Verbist, A. *et al.* (2007) Highly reliable and extremely stable SiGe micro-mirrors. Proceedings of the Micro-electromechanical Systems (MEMS), pp. 759–762.

38. Sangameswaran, S., De Coster, J., Linten, D. *et al.* (2008) ESD reliability issues in microelectromechanical systems (MEMS): A case study of micromirrors. Proceedings of the Electrical Overstress/Electrostatic Discharge (EOS/ESD) Symposium, pp. 249–257.

39. Voldman, S. and Gross, V. (1994) Scaling, optimization, and design considerations of electrostatic discharge protection circuits in CMOS technology. *Journal of Electrostatics*, **33** (3), 327–357.

40. Voldman, S. and Gerosa, G. (1994) Mixed voltage interface ESD protection circuits for advanced micro processors in shallow trench and LOCOS isolation CMOS technology. International Electron Device Meeting (IEDM) Technical Digest, Session 10.3.1, December 1994, pp. 277–281.

41. Voldman, S. (1993) Shallow trench isolation double-diode electrostatic discharge circuit and interaction with DRAM circuitry. *Journal of Electrostatics*, **31**, 237–262.

42. Voldman, S. (1999) Electrostatic discharge protection, scaling, and ion implantation in advanced semiconductor technologies, Invited Talk, Process Integration Issues/Technical Trends Session. Proceedings of the Ion Implantation Conference (I2CON), Napa, California.

43. Voldman, S. (1997) ESD robustness and scaling implications of aluminium and copper interconnects in advanced semiconductor technology. Proceedings of the Electrostatic Overstress/Electrostatic Discharge (EOS/ESD) Symposium, Sept. 1997, pp. 316–329.

44. Voldman, S. (1998) High current transmission line pulse characterization of aluminium and copper interconnects for advanced CMOS semiconductor technologies. Proceedings of the International Reliability Physics Symposium (IRPS), pp. 293–301.

45. Voldman, S., Morriseau, K., Hargrove, M. *et al.* (1999) High-current characterization of dual-damascene copper interconnects in SiO_2 and low-K inter-level dielectrics for advanced CMOS technologies. Proceedings of the IEEE International Reliability Physics Symposium (IRPS), pp. 144–153.

46. Voldman, S., Hui, D., Warriner, L. *et al.* (1999) Electrostatic discharge protection in silicon-on-insulator technology. Proceedings of the IEEE International silicon on insulator (SOI) Conference, pp. 68–72.

47. Voldman, S. (1999) Electrostatic discharge (ESD) protection in silicon-on-insulator (SOI) CMOS technology with aluminium and copper interconnects in advanced microprocessor semiconductor chips. Proceedings of the Electrical Overstress/Electrostatic Discharge (EOS/ESD) Symposium, pp. 105–115.

48. Russ, C., Gossner, H., Schulz, T. *et al.* (2005) ESD evaluation of emerging MUGFET technology. Proceedings of the Electrical Overstress/Electrostatic Discharge (EOS/ESD) Symposium, pp. 280–289.

49. Gossner, H., Russ, C., Siegelin, F. *et al.* (2006) Unique ESD failure mechanism in a MUGFET technology. International Electron Device Meeting (IEDM) Technical Digest, pp. 1–4.

50. Tremouilles, D. Thijs, S. Groeseneken, G. *et al.* (2007) Understanding the optimization of sub 45 nm FinFET devices for ESD applications. Proceedings of the Electrical Overstress/Electrostatic Discharge (EOS/ESD) Symposium, pp. 408–415.

51. Yeo, K.H., Suk, S.D., Li, M. *et al.* (2006) Gate all-around (GAA) twin silicon nanowire MOSFET (TSNWFET) with 15 nm length gate and 4nm radius nanowires. International Electron Device Meeting (IEDM) Technical Digest, pp. 1–4.

52. Liu, W., Liou, J.J., Chung, A. *et al.* (2009) Electrostatic discharge of Si nanowire field effect transistors. *IEEE Electron Device Letters*, **EDL-30** (9), 969–971.

53. Chen, J. and Voldman, S. (January 18 2011) Carbon nanotube diodes and electrostatic discharge circuits and methods. U.S. Patent No. 7,872,334.

54. Voldman, S. (1999) The impact of technology evolution and scaling on electrostatic discharge (ESD) protection on high-pin-count high-performance microprocessors, *International Solid-State Circuits*. Conference Proceedings, Session WA21, San Francisco, CA, Feb. 15–17, 1999.

55. Voldman, S. (2011) Nano ESD: Electrostatic Discharge in the Nanoelectronic Era, Chapter 15, in *Nanoelectronics: Nanowires, Molecular Electronics, and Nanodevices*, McGraw Hill, New York.

56. Voldman, S. (1998) The impact of MOSFET technology evolution and scaling on electrostatic discharge protection, *Review Paper. Micro-electronics Reliability*, **38**, 1649–1668.

57. Singh, N., Agarwal, A., Bera, L.K. *et al.* (2006) High performance fully depleted silicon nanowire (diameter < 5nm) gate-all-around CMOS devices. *IEEE Electron Device Letters*, **EDL-27** (5), 383–386.

Glossary

Air Ionizers: An electronic or nuclear device that generate ions from air to be used for dissipation of static charge, typically used in manufacturing and assembly environments.

Antistatic: Material or coating that prevents static buildup on worksurfaces, or materials. An antistatic agent is a compound used for treatment of materials or their surfaces in order to reduce or eliminate buildup of static electricity generally caused by the triboelectric effect.

Audits: Business processes review to verify conformance and compliance to ESD procedures and standards.

Cable Discharge Event (CDE): an electrostatic discharge event from a cable source.

Cassette Model: a test method whose source is a capacitor network with a 10 pF capacitor. This is also known as the Small Charge Model (SCM), and the Nintendo model.

Charged Board Event (CBE): a test method for evaluation of the charging of a packaged semiconductor chip mounted on a board, followed by a grounding process. The semiconductor chip is mounted on a board during this test procedure. The board is placed on an insulator during this test.

Charged Device Model (CDM): a test method for evaluation of the charging of a packaged semiconductor chip, followed by a grounding pin. The semiconductor chip is not socketed but placed on an insulator during the test.

Conductor: A material that allows free flow of electrons. An example of conductors includes metal materials such as copper and aluminum. A material whose conductivity exceeds that of insulators and semiconductors.

Electromagnetic Interference (EMI): An electromagnetic disturbance that affects an electrical circuit due to either electromagnetic induction or electromagnetic radiation emitted from an external source.

ESD Basics: From Semiconductor Manufacturing to Product Use, First Edition. Steven H. Voldman.
© 2012 John Wiley & Sons, Ltd. Published 2012 by John Wiley & Sons, Ltd.

Electromagnetic Pulse (EMP): A large electromagnetic burst event typically resulting from high energy or nuclear explosions. The resulting rapidly-changing electric fields and magnetic fields may couple with electrical/electronic systems to produce damaging current and voltage surges.

Electrical Overstress (EOS): an electrical event, of either over-voltage or over-current, that leads to electrical component or electronic system damage and failure.

Electromagnetic Compatibility (EMC): A branch of electrical sciences which studies the unintentional generation, propagation and reception of electromagnetic energy. Electromagnetic compatibility must address both the susceptibility of systems to electromagnetic interference, and the propagation of electromagnetic noise.

Electrostatic Discharge (ESD): Electrostatic discharge (ESD) is a subclass of electrical overstress and may cause immediate device failure, permanent parameter shifts and latent damage causing increased degradation rate.

Electrostatic Shielding: Shielding used in electronic systems to prevent the entry or penetration of electromagnetic noise.

Electrostatic Susceptibility: The sensitivity of a system to electromagnetic interference.

Equi-potential: A surface where all points on the surface are at the same electrical potential.

Equi-potential Bonding: A process where two objects are "bonded" whose electrostatic potential is the same to avoid electrostatic discharge from occurring.

ESD Control Program: A corporate program or process for addressing electrostatic discharge issues in manufacturing and handling in a corporation.

Field Induced Charging: Charging process initiated on an object after placement within an electric field. This is also known as Charging by Induction.

Human Body Model (HBM): A test method whose source is a RC network with a 100 pF capacitor and 1500 ohm series resistor.

Human Metal Model (HMM): A test method that applies a IEC 61000-4-2 pulse to a semiconductor chip; only external pins exposed to system level ports are tested. Source can be an ESD gun that satisfies the IEC 61000-4-2 standard.

Inductive Charging: A charging process that uses an electromagnetic field to transfer energy between two objects.

Insulator: A material whose conductivity is less than a conductor and semiconductor (less than 10^{-8} siemens per centimeter). Insulators are used to prevent flow of electrical current.

Integrated Circuit: An electrical circuit constructed from semiconductor processing where different electrical components are integrated on the same substrate or wafer.

Ionization: A method to generate ions from atoms. Ionization techniques include both electrical as well as nuclear sources.

Latchup: A process electrical failure occurs in a semiconductor component or power system where a parasitic pnpn (also known as a silicon controlled rectifier, thyristor, or Schockley diode) undergoes a high current/low voltage state. Latchup can lead to thermal failure and system destruction.

Latent Failure Mechanism: A failure mechanism where the damage created deviates from the untested or virgin device or system. A latent failure can be a yield or reliability issue.

Machine Model (MM): A test method whose source is a capacitor network with a 200 pF capacitor.

Semiconductor: A material whose conductivity is between a conductor and an insulator (in the range of 10^3 to 10^{-8} siemens per centimeter). Semiconductors are commonly used in integrated circuit component technology.

Small Charge Model (SCM): A test method whose source is a capacitor network with a 10 pF capacitor.

Socketed Device Model (SDM): A test method for evaluation of the charging of a packaged semiconductor chip, followed by a grounding pin. The semiconductor chip is socketed during the test.

Static Electricity: Electrical charge generated from charging processes that are sustained and accumulated on an object.

Surface Resistivity: The resistance of a material on its surface (as opposed to a bulk resistivity).

System Level IEC 61000-4-2: A system level test that applies a pulse to a system using an ESD gun.

Transmission Line Pulse (TLP): A test method that applies a rectangular pulse to a component (10 ns rise and fall time; 100 ns plateau).

Tribo-electric Charging: A method which generates charging through contact electrification. Contact electrification is when certain materials become electrically charged after they come into contact with another different material and are then separated (such as through rubbing). The polarity and strength of the charges produced differ according to the materials, surface roughness, temperature, strain, and other properties.

Tribo-electric Series: the ordering of materials according to their triboelectric behavior. Materials are often listed in order of the polarity of charge separation when they are touched with another object. A material towards the bottom of the series, when touched to a material near the top of the series, will attain a more negative charge, and vice versa. Tribo is from the Greek for "rubbing", $\tau\rho\acute{\iota}\beta\omega$ ($\tau\rho\iota\beta\acute{\eta}$: friction).

Very Fast Transmission Line Pulse (VF-TLP): a test method that applies a rectangular pulse to a component (1 ns rise and fall time; 10 ns plateau).

ESD Standards

ESD ASSOCIATION

ANSI/ESD S1.1-2006 Wrist Straps

ESD DSTM2.1- Garments

ANSI/ESD STM3.1 – 2006 Ionization

ANSI/ESD SP3.3-2006 Periodic Verification of Air Ionizers

ANSI/ESD STM4.1-2006 Worksurfaces – Resistance Measurements

ANSI/ESD STM3.1 – 2006 ESD Protective Worksurfaces – Charge Dissipation Characteristics

ANSI/ESD STM5.1 – 2007 Electrostatic Discharge Sensitivity Testing – Human Body Model (HBM) Component Level

ANSI/ESD STM5.1.1 – 2006 Human Body Model (HBM) and Machine Model (MM) Alternative Test Method: Supply Pin Ganging – Component Level

ANSI/ESD STM5.1.2 – 2006 Human Body Model (HBM) and Machine Model (MM) Alternative Test Method: Split Signal Pin – Component Level

ANSI/ESD S5.2 – 2006 Electrostatic Discharge Sensitivity Testing – Machine Model (MM) Component Level

ANSI/ESD S5.3.1 – 2009 Charged Device Model (CDM) - Component Level

ANSI/ESD SP5.3.2 – 2008 Electrostatic Discharge Sensitivity Testing – Socketed Device Model (SDM) Component Level

ESD Basics: From Semiconductor Manufacturing to Product Use, First Edition. Steven H. Voldman.
© 2012 John Wiley & Sons, Ltd. Published 2012 by John Wiley & Sons, Ltd.

ANSI/ESD STM5.5.1 – 2008 Electrostatic Discharge Sensitivity Testing – Transmission Line Pulse (TLP) Component Level

ANSI/ESD SP5.5.2 – 2007 Electrostatic Discharge Sensitivity Testing – Very Fast Transmission Line Pulse (VF-TLP) Component Level

ANSI/ESD SP6.1 – 2009 Grounding

ANSI/ESD S7.1 – 2005 Resistive Characterization of Materials – Floor Materials

ANSI/ESD S8.1 – 2007 Symbols – ESD Awareness

ANSI/ESD STM9.1-2006 Footwear – Resistive Characterization

ESD SP9.2-2003 Footwear – Foot Grounders Resistive Characterization

ANSI/ESD SP10.1-2007 Automatic Handling Equipment (AHE)

ANSI/ESD STM11.11-2006 Surface Resistance Measurement of Static Dissipative Planar Materials

ESD DSTM11.13-2009 Two Point Resistance Measurement

ANSI/ESD STM11.31-2006 Bags

ANSI/ESD STM12.1-2006 Seating–Resistive Measurements

ESD STM13.1-2000 Electrical Soldering/Desoldering Hand Tools

ANSI/ESD SP14.1 – System Level Electrostatic Discharge (ESD) Simulator Verification

ESD SP14.3-2009 System Level Electrostatic Discharge (ESD) Measurement of Cable Discharge Current

ANSI/ESD SP15.1-2005 In Use Resistance Testing of Gloves and Finger Cots

ANSI/ESD S20.20-2007 Protection of Electrical and Electronic Parts, Assemblies, and Equipment

ANSI/ESD STM97.1-2006 Floor Materials and Footwear – Resistance Measurements in Combination with A Person

DEPARTMENT OF DEFENSE

DOD HDBK 263 – Electrostatic Discharge Control Handbook for Protection of Electrical and Electronic Parts, Assemblies and Equipment.

DOD-STD-1686 – Electrostatic Discharge Control Program for Protection of Electrical and Electronic Parts, Assemblies and Equipment.

DOD-STD-2000-2A Part and Component Mounting for High Quality/High Reliability Soldered Electrical and Electronic Assembly.

DOD-STD-2000-3A Criteria for High Quality/High Reliability Soldering Technology.

DOD-STD-2000-4A General Purpose Soldering Requirements for Electrical and Electronic Equipment.

FED Test Method STD 101 - Method 4046 – Electrostatic Properties of Materials.

MILITARY STANDARDS

MIL-STD-454 Standard General Requirements for Electronic Equipment.

MIL-STD-785 – Reliability Program for System and Equipment Development and Production.

MIL-STD-883 - Method 3015-4 – Electrostatic Discharge Sensitivity Classification

MIL-STD-1686A – Electrostatic Discharge Control Program for Protection of Electrical and Electronic Parts, Assemblies and Equipment.

MIL-E-17555 – Electronic and Electrical Equipment, Accessories, and Provisioned Items (Repair Parts: Packaging of)

MIL-M-38510 – Microcircuits, General Specification for

MIL-D-81705 – Barrier Materials, Flexible, Electrostatic Free, Heat Sealable.

MIL-D-81997 – Pouches, Cushioned, Flexible, Electrostatic Free, Reclosable, Transparent

MIL-D-82646 – Plastic Film, Conductive, Heat Sealable, Flexible

MIL-D-82647 – Bags, Pouches, Conductive, Plastic, Heat Sealable, Flexible

IEC 801-2 – Electromagnetic Compatibility for Industrial Process Measurements and Control Equipment, Part 2: Electrostatic Discharge (ESD) Requirements.

EIA-541– Packaging Material Standards for ESD Sensitive Materials

JEDEC 108 – Distributor Requirements for Handling Electrostatic Discharge Sensitive (ESDS) Devices

Index

ESD Basics: From Semiconductor Manufacturing to Product Use, First Edition. Steven H. Voldman.
© 2012 John Wiley & Sons, Ltd. Published 2012 by John Wiley & Sons, Ltd.

Printed and bound by CPI Group (UK) Ltd, Croydon, CR0 4YY

27/10/2024

14580216-0002